HAVERFORD SCHOOL
PENNA. DEPT. OF EDUCATION
ACT 88
SCHOOL YEAR 1978-79

HAVERFORD SCHOOL
PENNA. DEPT. OF EDUCATION
ACT 88
SCHOOL YEAR

SECOND EDITION

LABORATORY MICROBIOLOGY

L. JACK BRADSHAW, Ph.D.

Professor of Biology
California State University, Fullerton,
Fullerton, California

W. B. SAUNDERS COMPANY • PHILADELPHIA • LONDON • TORONTO

W. B. Saunders Company: West Washington Square
Philadelphia, PA 19105

1 St. Anne's Road
Eastbourne, East Sussex BN21 3UN, England

1 Goldthorne Avenue
Toronto, Ontario M8Z 5T9, Canada

Listed here is the latest translated edition of this book together with the language of the translation and the publisher.

Spanish (2nd Edition) — El Manual Moderno, S.A.,
Mexico City, Mexico

Laboratory Microbiology   ISBN 0-7216-1906-1

© 1973 by W. B. Saunders Company. Copyright 1963 by W. B. Saunders Company. Copyright under the International Copyright Union. All rights reserved. This book is protected by copyright. No part of it may be reproduced, stored in a retrieval system, or transmitted in any form or by any means, electronic, mechanical, photocopying, recording, or otherwise, without written permission from the publisher. Made in the United States of America. Press of W. B. Saunders Company. Library of Congress catalog card number 72-95827.

Print No: 9   8   7   6

Preface to the Second Edition

In revising this book, the basic philosophy utilized was essentially unchanged from that of the first edition: to attempt to give the student a set of clear instructions which would make it possible for him to learn the basic techniques of experimental microbiology. Emphasis was placed on unequivocal understanding of the procedures or experiments to be performed as well as the theory behind them. The entire book was rewritten for updating purposes as far as the information content is concerned.

Thanks to the many helpful suggestions of users, various new features were incorporated, ranging from the addition of new experiments to the renaming of old ones. A set of "mini-biographies" of each organism at the time it is first introduced has been included to give the student something more than just a name to hold on to. Several new illustrations were added to assist the instructor in emphasizing a particular technique graphically. The new experiments in applied microbiology were added in response to numerous requests from instructors using the book. Finally, several optional experiments are included which involve spectroscopy, in the hope that this technique will prove useful and informative. It is very likely that a greater use of this and other similar quantitating devices will become a routine adjunct to the beginning microbiology laboratory of the future, as it is already in many of them.

I would like to acknowledge the helpful suggestions that have come from many users of *Laboratory Microbiology* and particularly to my former colleague, Dr. John Lewis. I would also like to express my sincere appreciation to Judy Smith for her help in preparing the manuscript for this revision. Finally, I would like to thank Julie Knapp for her careful proofreading and helpful suggestions during the preparation of the manuscript.

L. JACK BRADSHAW

Fullerton, California

Preface
to the First Edition

As is probably the case with most textbooks, this one originated as an attempt to solve some local problems in the teaching of laboratory microbiology. In this case, the principal problems fell into two main categories: the student needed immediate access to more specific information bearing upon the experimental work to be performed, in order to develop an understanding of the ideas behind the procedures; and the instructor needed a more positive "quiz-ahead" device to insure ample pre-reading of the experiments by the students. Both problems were resolved by the same idea — that of writing text material in sufficient detail that the student could be held responsible for a prior knowledge of the experiments to be performed. This book is therefore somewhat of a mixture of the textbook and the traditional laboratory manual.

Also growing out of this general thinking was the realization that most laboratory manuals assume a great deal of prior knowledge on the part of the student, an assumption that is often mistaken. For example, much of the study of bacterial cytology is based upon principles commonly taught in a course in micro-technique, which most beginning students in general microbiology have not taken. Therefore the understanding of the student concerning the true nature and purposes of staining is usually far short of what it should be for him to appreciate that portion of the microbiology laboratory.

The general pattern followed in this laboratory textbook is that of a flowing discussion on the theory behind a given experiment or group of experiments, interrupted at logical points with instructions for performing the experiment. These instructions have been purposely written in a discursive fashion rather than enumerating each step as is usually done in most laboratory manuals. The author believes this type of approach compels the student to read the instructions critically and thereby gives him an additional groundwork of information before he comes to the actual laboratory work. At the same time, students should be encouraged to draw up their own itemized "plan of attack" before coming to the laboratory, enumerating the steps to be followed, so the experiment can be performed the most efficient way possible. Usually, two or more experiments are performed during the same laboratory period, and a thorough study of the procedures involved can result in a specific plan that will combine much of the work to be done. For example, while an agar tall is melting during one experiment, some steps of the procedure for the second one could be performed. In this way several experiments can be blended into a single set of step-by-step instructions. In our laboratory we hand out blank monthly calendars on which the student can jot down brief daily instructions for his use. This procedure has virtually eliminated the forgetting of experiments that require extended incubation and observation.

In recognition of the fact that prerequisite courses for general microbiology vary from one institution to another, the chemical explanations are kept at a fairly descriptive level throughout the text. It is easier for an instructor to amplify an explanation to his class than it is to "tone down" an explanation appearing in a text. If a full year of college chemistry (or more) is a prerequisite for the microbiology course, some additional explanation of the chemical activities of the microorganisms is undoubtedly called for.

It should be mentioned that most of the experiments in this book are similar to those considered "standard" in most laboratory manuals; other experiments are modifications or perhaps innovations developed in the author's laboratory. In most cases general experiments are included, with very little "applied microbiology" being utilized. In the choice of experiments, great emphasis has been placed upon experiments that always produce correct results, since the beginning student tends to become confused when a given procedure does not proceed as expected. A good deal of detail is included in instructions for the preparation of media, dyes and reagents, because student help is frequently used in their preparation.

A modified key taken from *Bergey's Manual of Determinative Bacteriology* (7th Ed.) is included in this text. This has been found to be of great assistance in introducing the student to the keying-out of an unknown bacterium without incurring the work usually encountered with direct use of the manual. This modified key is not intended to be comprehensive or an actual substitute for *Bergey's Manual,* but rather an introduction to the use of this technique. The key is limited to those organisms whose specific characteristics are listed following it; in practice this system works rather well.

I would like to express my gratitude to my most understanding family for the many times I was not able to be with them because of work on this manuscript, and especially to my wife for her valuable assistance.

L. JACK BRADSHAW

Contents

CHAPTER 1

INTRODUCTION TO THE MICROBIOLOGY LABORATORY 1

The Purpose of Laboratory Study 1
Reporting Results 3
Introduction to the Microscope 3
    Experiment 1.1, Introduction to the Use of the
    Microscope 7
Introduction to Some Common Microbiological Apparatus 7
    Culture Tubes 7
    Petri Dishes 9
    Bacteriological Pipets 9
    Experiment 1.2, The Use of Pipets, Indicators,
    and Buffers 14
    Bacteriological Incubators and Water Baths 15
    Bacteriological Sterilizers 15
    The Inoculating Loop and Needle 16
    Bacteriological Slides 16
Introduction to Some Microbiological Techniques 16
    The Isolation of a Pure Culture of Bacteria 18
    Experiment 1.3, Isolation Techniques and Use of
    Petri Dish Cultures 19
    Streak Plate Method of Isolation 19
    Isolation of Bacteria with a Bent Glass Rod 22
    Sectoral or Zone Growth 23
    Isolation of Microbial Growth from a Cotton Swab 23
    Sterile Technique 24
    Experiment 1.4, The Omnipresence of
    Microorganisms in the Environment and the
    Necessity for Aseptic Technique 24
    Anaerobic Culture Methods 25
    Experiment 1.5, The Cultivation of Anaerobic Organisms ... 27

viii CONTENTS

CHAPTER 2

CYTOLOGY OF MICROORGANISMS 28

Experiment 2.1, Smear Preparations and Fixation 30
Experiment 2.2, Negative Staining of Microorganisms 33
Experiment 2.3, Motility Studies on Microorganisms 34
Experiment 2.4, The Demonstration of Fat Vacuoles 36
Experiment 2.5, The Demonstration of Metachromatic Granules (Volutin) 37
Experiment 2.6, The Gram Stain 40
Experiment 2.7, The Acid-Fast Stain 42
Experiment 2.8, Demonstration of Bacterial Endospores ... 44
Experiment 2.9, Demonstration of Bacterial Capsules 45
Experiment 2.10, The Flagella Stain 46
Experiment 2.11, Demonstration of the Bacterial Cell Wall .. 47
Experiment 2.12, Demonstration of the Nucleus and General Morphology of Yeasts 48
Experiment 2.13, A Study of Mold Morphology 50
Preparation of a Mold Microculture 51
Observation of Molds from the Air 51
Experiment 2.14, The Study of Protozoan Morphology 52
Experiment 2.15, Cytological Characterization of an Unknown Bacterium 54

CHAPTER 3

NUTRITION OF MICROORGANISMS 55

Basic Growth Requirements 55
Experiment 3.1, Determination of Minimal Growth Requirements 57
Adaptations of Media to Studies of Microorganisms 57
Experiment 3.2, Selective and Differential Media 58
Experiment 3.3, The Utilization of Unusual Sources of Nitrogen and Carbon by Microorganisms 59

CHAPTER 4

METABOLISM OF MICROORGANISMS 60

Exoenzymes of Microorganisms 60
Experiment 4.1, The Hydrolysis of Starch 62
Experiment 4.2, Hydrolysis of Casein 63
Experiment 4.3, The Hydrolysis of Gelatin 65
Experiment 4.4, Hemolysis Production 67
Experiment 4.5, Coagulase Production (Demonstration) 67
Endoenzymes of Microorganisms 68
Experiment 4.6, Alcohol Production by Yeasts 70
Experiment 4.7, Lactic Acid Production by Bacteria 71

Experiment 4.8, Fermentation of Carbohydrates by
  Microorganisms 73
Experiment 4.9, Reduction of Methylene Blue 75
Experiment 4.10, Reduction of Litmus Milk 77
Experiment 4.11, Reduction of Nitrates 78
Experiment 4.12, The Production of Hydrogen
  Sulfide by Microorganisms 79
Experiment 4.13, Demonstration of the
  Production of Indole 80
Experiment 4.14, The MR-VP Procedure 81
Experiment 4.15, The Detection of Catalase in Bacteria 82

CHAPTER 5

DESTRUCTION AND INHIBITION OF MICROORGANISMS 83

Environmental and Physical Effects 83
  Experiment 5.1, The Effect of Hydrogen Ion
    Concentration (pH) upon Microorganisms 84
  Experiment 5.2, The Effect of Osmotic Pressure on
    Microorganisms 86
  Experiment 5.3, The Effect of Ultraviolet
    Radiation on Microorganisms 88
  Experiment 5.4, The Effect of Heat on Microbial Survival .. 89
Planned Destruction of Microorganisms 91
  Experiment 5.5, Demonstration of Molecular
    Diffusion in Agar 92
  Experiment 5.6, The Selective Action of Crystal Violet 93
  Experiment 5.7, The Oligodynamic Action of Heavy Metals . 94
  Experiment 5.8, The Filter Paper Disc Method of
    Evaluating Proprietary (Commercial) Antiseptics 96
  Experiment 5.9, The Evaluation of Antibiotics by the
    Filter Paper Disc Method 97
  Experiment 5.10, Determination of the Rate of Action of
    Antiseptics With and Without Organic Matter Present 98
  Experiment 5.11, Some Observations on Skin
    Cleanliness and Skin Antisepsis 99

CHAPTER 6

IDENTIFICATION OF UNKNOWN BACTERIA 101

  Experiment 6.1, Identification of an Unknown Bacterium . 105
Abridged Key for Identification of Some Common Bacteria 106
Short Key for the Family Determination of Unknown Bacteria .. 112
Specific Characteristics of Selected Bacteria 114

CHAPTER 7
ECOLOGY OF MICROORGANISMS 118

Experiment 7.1, Bacterial Commensalism 119
Experiment 7.2, Experimental Antibiosis 120
Experiment 7.3, Antibiosis in Soil.................... 120
Experiment 7.4, Bacterial Synergism................... 122
Experiment 7.5, Symbiotic Nitrogen Fixing Bacteria 123
Experiment 7.6, The Relationship of Microorganisms to Dental Decay 124
Experiment 7.7, Normal Flora of the Human Skin 126
Experiment 7.8, Normal Flora of the Human Throat 127

CHAPTER 8
APPLIED MICROBIOLOGY 128

Experiment 8.1, The Bacteriological Analysis of Water 129
Experiment 8.2, The Bacterial Growth Curve 132
Experiment 8.3, The Plate Count of Bacteria in Milk 136
Experiment 8.4, The Effect of Temperature on Microbial Growth in Milk 137

CHAPTER 9
SELECTED MEDICAL MICROBIOLOGY LABORATORY PROCEDURES 139

Experiment 9.1, Isolation and Identification of an Enteric Pathogen 141
Experiment 9.2, Selective Medium for the Isolation of Pathogenic Staphylococci 143
Experiment 9.3, The Agglutination Reaction 145
Experiment 9.4, The Morphology of Blood Cells, Erythrocyte Blood Typing and Differential Count 147

EXPERIMENT WORK SHEETS 151

APPENDIX 291

GENERAL MICROBIOLOGY LABORATORY INSTRUCTIONS

STUDENT EQUIPMENT

The following items of equipment should be available to each student:

One box of glass microscope slides (approximately 10).
One concave slide and one deep well slide.
One inoculating needle holder, one straight nichrome wire (4 in.) and one nichrome loop (3 in.)
One bent glass rod.
One slide holder.
Six coverslips.
A large handkerchief or small dish towel for drying slides.
Lens paper.
One tin can for boiling test tubes.

The following items should be available to students in the laboratory for general use:

Bunsen burners.
Tripods and wire gauze to support cans of water for boiling.
Centigrade thermometers (0° - 100°).
Sterile Petri dishes.
Sterile pipets (1.0 and 10.0 ml. sizes).
Test tube racks or supports.
Bottles of sterile nutrient agar for plating (prescription bottles work well).
Incubator (37°C.).

GENERAL RULES OF LABORATORY CONDUCT

Although specific differences in rules exist from laboratory to laboratory, there are a few rules of conduct that are more or less universal in their applicability:

1. Eating or drinking is prohibited in the laboratory at all times.

2. Avoid putting any objects (such as pencils, fingers, etc.) in your mouth while working in the laboratory.
3. Wash your hands with soap and water at the conclusion of each laboratory period.
4. If a living culture of microorganisms is spilled, notify the instructor in charge of the laboratory immediately.
5. If you are injured in any way (burning is the most common type of injury sustained in this laboratory), notify the instructor immediately.
6. Make sure that all gas, water, and electrical appliances at your table are turned off before you leave.
7. Unless otherwise specified, all cultures are to be grown in the 37°C. incubator. If room temperature is called for, incubate in your drawer.
8. When melting a container of agar by boiling, make very sure that the cap is loosened slightly; otherwise the bottle may explode.
9. Be sure you wipe off the immersion oil before putting your microscope away.

CHAPTER 1

INTRODUCTION TO THE MICROBIOLOGY LABORATORY

THE PURPOSE OF LABORATORY STUDY

Too many times the student views laboratory study as a tiresome assignment to be endured and the laboratory as a place where one only copies down results previously known. Because of this attitude, many students completely lose sight of the purpose of laboratory study. This is most unfortunate, because, given the proper attitude and a reasonable amount of intellectual curiosity, the student should find laboratory study one of the most fascinating parts of his academic experience. An observer can usually tell when a good laboratory is in operation; the students are not anxious to leave and often spend many more than the scheduled hours in the laboratory. It is hoped that each student will be able to view the microbiology laboratory in this manner. Science, by its very nature, is a product of laboratory effort and thus one can say that real science is in the laboratory.

At this point, it might be well to call attention to a few simple but pertinent facts regarding the usefulness of laboratory study to the student. Undoubtedly, the main purpose of laboratory manipulation is to develop concepts learned from reading books and journals. In science, as in other fields of endeavor, there is often a large gap between book knowledge and applied knowledge. This is particularly true of modern microbiology, in which new and fascinating laboratory techniques and ideas are encountered almost daily in the journals. Unfortunately, many of these new techniques are predicated on the availability of excellent laboratory facilities to the experimenter. When the student is able to read about a certain phenomenon and then is able to follow this up by learning "how to do it" in a physical sense, learning is reinforced and a more realistic form of knowledge is acquired. It is quite true that many of the experiments set up in a student laboratory result in answers obvious to the experimenter before he ever begins, but it is by no means certain that the student can experimentally obtain this answer because of his limited skill in laboratory techniques. You may know the result you're supposed to obtain, but you may not always be able to get it when performing the experiment yourself. Often, even the most obvious experiment requires several repetitions before skill adequate to produce the desired results is acquired. Therefore, the "simple" experiment may become not so simple when viewed in this light.

A student who wishes to gain a full measure of return from his efforts in the laboratory should be prepared to repeat experiments that are not successful the first time. Satisfaction with a sloppy result only works to the detriment of the student. Perhaps it should be pointed out that there will be times when the desired results cannot be achieved even through repetition, simply because in dealing with living things the student is not working with unvarying entities. It is hoped that the experimental organisms provided are of the type designated, but often, particularly in microbiology, the student is plagued by mutative changes in these organisms, so that many times they no longer behave as originally anticipated. When this happens, it is possible for the experimenter to achieve the same incorrect result consistently despite repeated attempts. This often discourages students who hope to achieve ideal results (that is, predictable results). Herein lies one of the greatest challenges in biological experimentation — the student deals with fluid and pliable systems that usually remain predictable under specified conditions but are capable of unexpected variation. It is hoped that this idea will help to show the student that sometimes how he achieves the result is more important than the result itself. Careful attention to technique is essential.

A primary feature of the microbiology laboratory which tends to set it apart from many others is the "living" nature of virtually every experiment. Rarely will you be looking at preserved specimens or parts of organisms. The microorganisms employed are "alive and kicking," and as a result, certain rules of procedure must be adhered to very strictly. Although the vast majority of the microorganisms are harmless to you, some obviously are not. For this reason, you will be taught techniques which will apply to all microorganisms, whether or not they are disease-causing, so that your personal safety in the laboratory will be assured. If you develop an automatic or reflex attitude toward these basic techniques, you need have nothing to fear from any of your experimental subjects.

One of the prime goals of those who conduct a science laboratory is to give the student added incentive to study through interest developed in the laboratory. Most of us find it more interesting in the long run to observe the results of our own efforts rather than to read about the same results achieved by others. It is hoped that the experiments selected for this book have been set up to provide a maximum of understanding and interest. Let the student be assured that there is so much material that must be covered in the time allotted that there is no time whatsoever for work of the type that simply consumes time. Every experiment has been selected from many possible experiments that could have been used, and it is hoped that each one will teach the student something new and interesting.

One of the most important factors in laboratory study, from the student's point of view, is the attitude that he carries into the laboratory. If the student is interested in learning and is willing to put in both time and effort, he will find it one of the most pleasant and rewarding experiences of his academic career. Carrying this attitude from the beginning, the student will find that the microbiology laboratory is an excellent stimulus to the imaginative mind. There are many experiments that could be performed that are not included in these laboratory instructions. If you want to perform experiments that are products of your own imagination, consult with the instructor, and the proper facilities probably will be made available.

REPORTING RESULTS

As in any experimental situation where an objective is being sought, it is necessary to report the experiment in an organized fashion. Experience has shown that memory is an exceedingly poor repository for experimental observations; the only safe method is to write them down. There are many different ways of doing this, but ultimately the objective is always the same — to record the observations in an organized fashion and in the briefest possible manner. The experimental scientist ordinarily records results in a daily log sheet and then compiles these results into a publication or paper that condenses this record into a short but meaningful essay. In the laboratory you will report your results in the following form when appropriate (report sheets are provided in the book):

(a) *The statement of purpose* section should consist of one or two sentences that describe the principal purpose of the experiment to be performed.

(b) The *data* section should include no editorializing but simply a statement of results obtained from the experiment without further comment. Tables or graphs are frequently useful here; sometimes drawings are the only data.

(c) The *discussion and conclusions* section certainly should be the most meaningful part of the entire report. In this section you discuss the results obtained in light of the objectives with which you began. You will be expected to criticize the results you have obtained and the methods employed. This section will indicate whether or not you have fully understood the experiment. The final part of this section should be a numerical listing of the pertinent conclusions (if any) that you have reached as a result of your work. These conclusions should be brief, simple statements.

This, then, is the general pattern that is followed in most scientific publications; these reports will serve as practice in this technique. The results of the laboratory work will be kept by the student in a manner prescribed by his instructor. All records of results must be written in the laboratory so that they are in a legible and readable condition any time the instructor requests them. There will be periodic individual *unannounced* checks on your daily progress.

INTRODUCTION TO THE MICROSCOPE

One of the most useful tools of the microbiologist is the optical microscope. Although this instrument is used perhaps less than most people

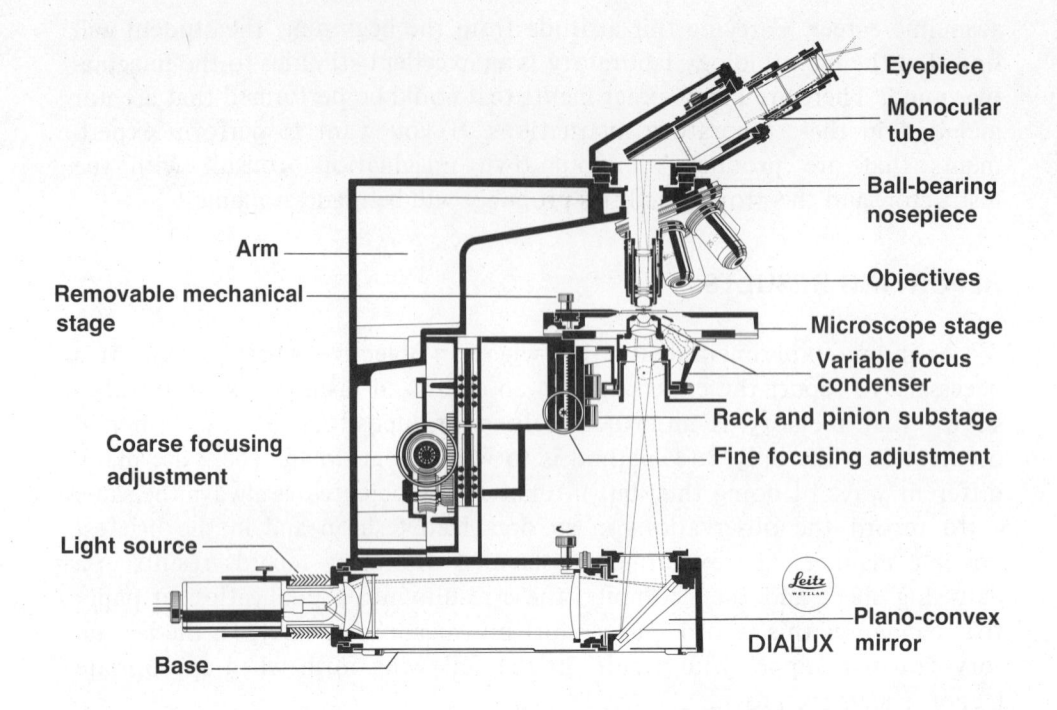

Figure 1-1 Typical working parts of the optical microscope. (Courtesy E. Leitz, Inc., Rockleigh, N.J.)

realize, it still is one of the most important instruments of study in microbiology and is certainly indispensable for certain types of work. All of the living forms with which you will be dealing are invisible to the naked eye, and a thorough familiarity with the microscope is essential at the very beginning of any laboratory course in microbiology. You will be expected to know all the major parts of the microscope and the function of each part.

Microscopes are considered either simple or compound, according to the number of lenses employed. Simple microscopes utilize single lenses, and compound microscopes utilize two or more lenses. Your microscopes are of the latter type, employing essentially two lens systems — the *ocular* or *eye piece,* and the *objective* lens. These two lenses are separated by a tube at such a distance that the eyepiece magnifies the image produced by the objective lens. In other words, the focal point of the objective lens is at some determined point within the tube, and the focal point of the ocular is set so that the ocular magnifies the image formed by the objective lens. The net result is a great enlargement of the object by the combined effects of the two lenses. Most microscopes used in microbiology require more than one type of objective lens since the magnifications needed differ from one type of experiment to another. Your microscopes are provided with three objective lenses, each giving a different magnification. The low power objective magnifies an object 10 times. The high-dry objective magnifies the object approximately 45 times, and the oil immersion objective magnifies an object 97 times. When combined with the ocular, which magnifies 10 times, a practical magnification is achieved in each of the objective lenses of 100, 450, and 970 times respectively. The term *magnification* means the enlargement of the linear diameter of an object.

Sheer magnification is not the only consideration the student is concerned with in the study of microorganisms. Undoubtedly the most important single theoretical consideration in the use of the microscope is a factor called *resolving power*. The resolving power of a microscope is a mathematical expression that denotes the ability of the lenses to distinguish detail clearly. One might say that the lens system loses its resolution of a given object when it can no longer separate two closely placed points or when the two points appear as one. Therefore, if a microscope cannot resolve an object, additional magnification is of no value, since it simply makes the blur larger. For all practical purposes, the limit of resolution of your microscope is approximately 0.2μ.

A consideration of the details of operation of your microscope shows that operation of this instrument is based on the adjustment of several variables. You have two controls that move the optical system up and down, which serves to bring the combined lens system into proper focus. The *coarse adjustment* moves the tube rather rapidly, and the *fine adjustment* moves the tube very slowly. The coarse adjustment is capable of moving the *barrel* or *tube* of the microscope up and down throughout the entire length of the gears. The fine adjustment can only move through a limited number of turns, at which time it comes to a stop. Therefore, prior to use, the fine adjustment should be set in the *middle* of its course of travel so that you have a maximum of travel with it in either direction. Many microscopes are equipped with *mechanical stages*. These devices grip the microscope slide and enable very precise regulation of the movement of the slide. Sometimes it is necessary to observe without using the mechanical stage, in which case it can be very simply removed, as will be demonstrated by the instructor. Another variable concerns the amount of light entering the lens system, controlled largely by the *iris diaphragm*. When the diaphragm is closed down tightly, virtually no light is permitted to pass through; when it is fully open, it permits entry of a maximum amount of light. There is sometimes an adjustment on the *substage condenser,* which also has some influence on the amount and type of light entering. This will be discussed by the instructor. In general, most of your work should be done utilizing the maximum amount of light (iris diaphragm open wide).

A very important aspect of the operation of your microscope involves the proper use of the adjustment knobs in bringing the object to be studied into proper focus. Two of the objective lenses on your microscope can travel down onto the glass slide, which might result in damage to both lens and slide. It is, therefore, imperative that the student learn the proper method for focusing a microscope at the beginning. This method involves using the coarse adjustment to move the objective lens down as close to the slide as possible while watching from the side to make sure the slide and lens do not touch. *Never look through the eyepiece when turning the adjustment knobs downward.* When the lens is nearly touching the glass slide, then and only then look through the ocular and begin focusing *upward* until you come into proper focus. The low power objective will not touch the glass when traveling to the bottom of its run, but both the high-dry and oil-immersion lenses will. Unfortunately, these are the most expensive of the objective lenses and are often subject to damage if pushed into the glass of the slide with any degree of force. Other details of the operation of your instrument will be given to you by your instructor.

Some of the common problems that beset beginning students while using the microscope should be mentioned at this time. There is an axiom in the field of electricity that states, "Always look for a blown fuse first in any electrical device that doesn't work." The lesson is simply that there are a few common recurring problems that generally constitute most of the difficulties in electricity or microscopy. Undoubtedly *the* most common problem is that of dirty lenses, both ocular and objective. All lenses should be cleaned with lens paper *each day* before use. The most effective method for cleaning lenses is the same as that employed for cleaning eye glasses. The student simply blows his breath on the lens to deposit a layer of condensation and wipes it off with the lens paper. Do not use handkerchiefs or other materials of this sort, since they usually make the lens dirtier. The eyepiece usually lifts out, and the "under" lens can be cleaned in the same way as the lens on the upper side. The objective lenses can ordinarily only be cleaned on the exposed surface, and in this case it is best done by taking a little tap water on your finger and moistening the lens surface, and then wiping it dry with lens paper. If this procedure still leaves the lens dirty, grease or oil may be the cause, in which case a small amount of xylene put on one corner of your lens paper and wiped briefly over the lens should remove the offending materials. Follow this with a brisk rubbing, using the dry part of your lens paper to remove the xylene. This solvent should be used quite sparingly, since in some microscopes the cement that holds the objective lenses in place is soluble in xylene, which could cause the lens to fall from its mounting. If you cannot see a subject through your microscope, the first thing you should look for is a dirty lens.

Another common error is the use of the high-dry lens instead of the oil-immersion lens. Since these two lenses are approximately the same length and look very similar, students sometimes use a high-dry lens with a drop of oil and wonder why the field of view is obscured. Make sure that you use the proper lens.

Yet another common mistake involves the proper centering of the lens to be used. The lens is not ready for use until it "clicks" into place under the barrel or tube of the microscope. Make sure that it is firmly snapped into place.

A common difficulty concerns improper illumination. On microscopes with fixed illumination, this does not present nearly so much a problem as it does with a mirror-type microscope. In general, however, your problems with illumination will be a matter of using too much illumination under certain circumstances, the field becoming so bright that you cannot see the object.

Special mention should be made regarding the cleaning of the oil immersion objective when you are through for the day. Wipe this lens with clean lens paper until no further evidence of oil is obtained. *Do not* use xylene or any other such solvent in cleaning the oil from this lens.

Many microscopes are *parfocal* — that is, when one lens is in focus, the other two lenses will also be at the proper focal length if they are put into position without changing any adjustments. In practice this means that you may come into clear focus under low power (which is the easiest lens to bring into focus) and then, without moving anything else, by simply switching lenses to the one desired, you should be within one half turn (one way or another) with the fine adjustment of the proper focus.

EXPERIMENT 1-1

Introduction to the Use of the Microscope

MATERIALS

1. A strand of hair.
2. A piece of paper with a torn edge.
3. Concentrated sodium chloride solution.
4. Fibers of cotton.

PROCEDURE: Examine all materials listed above under the low power objective. It is easier to hold the hair and cotton fibers securely under the microscope if you put them in the middle of a drop of water on a slide. This prevents them from blowing off the slide during the observation. The concentrated sodium chloride solution should be allowed to dry on the slide before observing. This causes the salt to form large crystals.

Repeat your observations using the high-dry objective. During observation with this objective, place a cover slip over each of your wet preparations to protect the lens from water. This is known as a <u>wet mount</u>. Make sketches under both low and high power of each different subject. Be sure to make your drawings definite or positive in outline. It is most important to pay attention to detail and actually to draw each subject in the field of view. Do not simply put down a mass of dots or lines from memory.

INTRODUCTION TO SOME COMMON MICROBIOLOGICAL APPARATUS

Culture Tubes

In the microbiological laboratory, just as in the chemistry laboratory, the common test tube finds a great deal of use. In microbiology, test tubes are used principally for growing microorganisms in artificial culture. This is somewhat analogous to the growing of animals in a cage or plants on a plot of ground. A *nutrient solution* is provided for organisms in the test tube. The organisms are then planted and growth takes place in this solution. A nutrient solution contains all of the necessary substances for the growth of the organism being studied. Experiments are included later that demonstrate this point more specifically. At this time, suffice it to say that the materials in the nutrient solution are analogous to the food on your dinner plate, and the same thing happens to the nutrient solution that happens to your food. These nutrient materials are converted into the living substance of the bacteria just as we grow because of the food we eat. Bacteria have no mouth parts, and all food substances diffuse through the membrane of the cell and must therefore be in solution before they can be usable.

Several different kinds of test tubes are used in microbiology. Traditionally, microbiological test tubes are plain and with no lip. They are stoppered with a plug of cotton. Many laboratories continue to use this type of growth tube. In other laboratories several modifications are found. Instead of cotton, a closure made of stainless steel called the *Morton test tube closure* may be employed. This is a sleevelike cap that fits firmly over the top of the test tube and serves the same purpose as a cotton plug. It has many advantages over the cotton plug aside from the fact that it is easy to put on. Some institutions use a rubber or plastic closure similar in idea to the Morton closure. The rubber or plastic closure, however, is heat sensitive, and many times, through improper use, students ruin it by overheating. The mainstay of many test tube collections probably is the *screw cap* test tube. It is threaded at the top and a plastic cap screwed on to it.

A common practice in sterile technique in the laboratory is the flaming of the neck of the tube with the Bunsen burner to destroy microorganisms that may have accumulated there before introducing the desired organisms into the test tube. Flaming should be adequate to kill microorganisms but not so much as to risk burning one's hands. One to two seconds is ordinarily adequate. Remember that glass holds its heat for a long time, so do not touch the flamed end of the tube with your hands or fingers for several minutes after this process.

A modification of the test tubes already described is the *Durham tube*. This consists of a small test tube placed inverted inside of a large test tube. This serves to collect and make visible any gas released by microorganisms during their growth. If the student wishes to measure the quantity of gas released, he must use another piece of apparatus known as the *Smith tube*. This essentially is a test tube in which the bottom is curved in the shape of a U. Most of the gas produced by microorganisms is collected in the side arm of the U-tube and measured volumetrically.

It is often desirable to use nutrient solutions that will harden or solidify as well as those that are in liquid form. The usual solidifying agent is an extract of seaweed called *agar*. Agar will harden a solution into a clear jell similar to ordinary gelatin, but it is different from gelatin in certain respects that make it much more suitable for use in the microbiology laboratory. For example, it is non-nutritive for most organisms, whereas gelatin (a protein) is utilized by many organisms as a nutrient. When gelatin is used as food, it is converted into liquid by the microorganisms, and the solidifying properties that it originally had are destroyed. Agar, usually not utilized as food, will remain solid.

Another advantage of agar results from its melting and solidifying temperatures. Incubation of most microorganisms is conducted at the melting temperature of gelatin (25°C.). A medium solidified by gelatin will therefore become fluid. Agar remains firm until very close to its boiling point, about 100°C. Once agar has melted it will remain fluid to approximately 40°C., just a little warmer than your body temperature. Both agar and gelatin form a clear transparent material when hard, thus making it possible to see through a reasonably thin layer of this medium.

Microorganisms can be grown conveniently in test tubes of nutrient solution containing agar (nutrient agar) if a suitable amount of nutrient surface is provided beforehand. This can be achieved by melting the agar in the test tube and then allowing it to harden in a slanted position. This is known as an *agar slant*. Sometimes, for various reasons, it is preferable to have the tube full of agar but not slanted. This is referred to as an *agar tall* or an *agar deep*. It should be pointed out that these terms do not indicate the character of the nutrient medium in any way. Therefore the usual practice is to precede such names with the name of the medium it contains, for example, *Sabouraud* agar slant, *nutrient* agar slant, and *E.M.B.* agar tall.

Petri Dishes

Petri dishes are made of glass or plastic and provide a place in which to grow microorganisms in artificial culture. The Petri dish differs from the test tube by providing a much broader surface on which to work. The Petri dish consists of two dishlike parts that look alike except that one is slightly larger and fits over the smaller, forming a cap. The usual Petri dish culture contains about 15 ml. of solid medium (agar-containing medium). The amount of medium in one agar tall is ordinarily approximately that quantity. You will pour agar into Petri dishes from a large stock bottle in later exercises, and you should attempt to use about 10-15 ml. each time. As a rule, once the agar has hardened in a Petri dish, the dish can be inverted without the agar dropping out. In most cases you will incubate Petri dish cultures *upside down*. The reason for this will be discussed later. Petri dishes will be provided, sterilized and ready to use.

Bacteriological Pipets

The bacteriological pipet is an exceedingly useful glass instrument that is used to deliver specified quantities of fluid from one container to another. Since it is ordinarily required that this transfer be made under *aseptic* conditions, the pipets provided are sterilized so that a solution may be transferred from one sterile container to another without loss of sterility.

INTRODUCTION TO THE MICROBIOLOGY LABORATORY

The word *asepsis,* strictly speaking, means freedom from *pathogenic,* or *disease-causing,* microorganisms. In this sense of the word, it is properly used in a purely medical situation such as would be found in surgical technique. However, it has become common parlance in the microbiology laboratory to use the phrase "aseptic conditions" or "aseptic technique" analogous for what more properly should be called *sterile technique.* Sterile technique provides for the transfer of materials without the introduction of any unwanted form of life. The term *sterile* means *devoid of all life.* In the usual bacteriological situation the student transfers living cultures from one container to another but wishes to avoid the introduction of unwanted organisms from either the atmosphere or objects in the area. Such organisms are referred to as *contaminants.* Therefore the term sterile technique is perhaps more correct for this, although aseptic technique in practice is used commonly.

A bacteriological pipet is essentially a piece of glass tubing that has been calibrated and drawn out to a capillary-type tip. At the other end the pipet has a mouthpiece where suction is applied to draw the fluid into the pipet. By proper control with the index finger, you can deliver any quantity you desire within the range of the pipet. You will be provided with pipets of two sizes: (a) The 1 ml. pipet draws a total of 1 ml. with visible graduations of 0.1 ml. or 0.01 ml., and is used whenever 1 ml. quantities (or less) are called

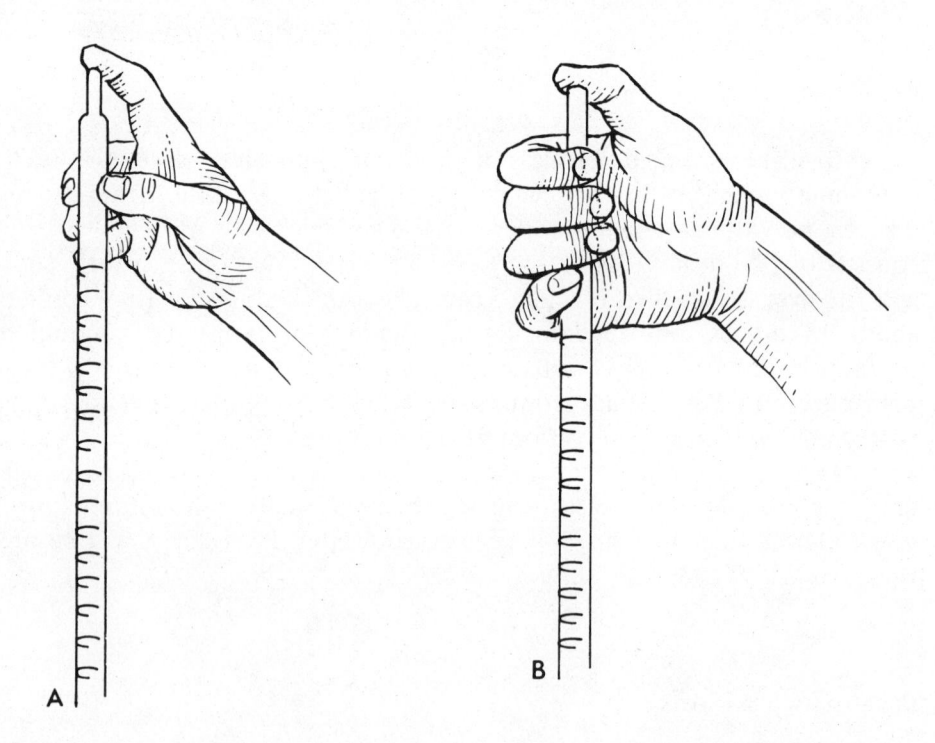

Figure 1-2 The drawings above illustrate correct and incorrect ways to hold a bacteriological pipet. The incorrect method (right) does not provide sensitive control of fluid flow as compared to the method shown on the left. Note the wide spread between the index finger and the other fingers in the latter. This hold affords maximum control of flow and stability of "aiming" the pipet.

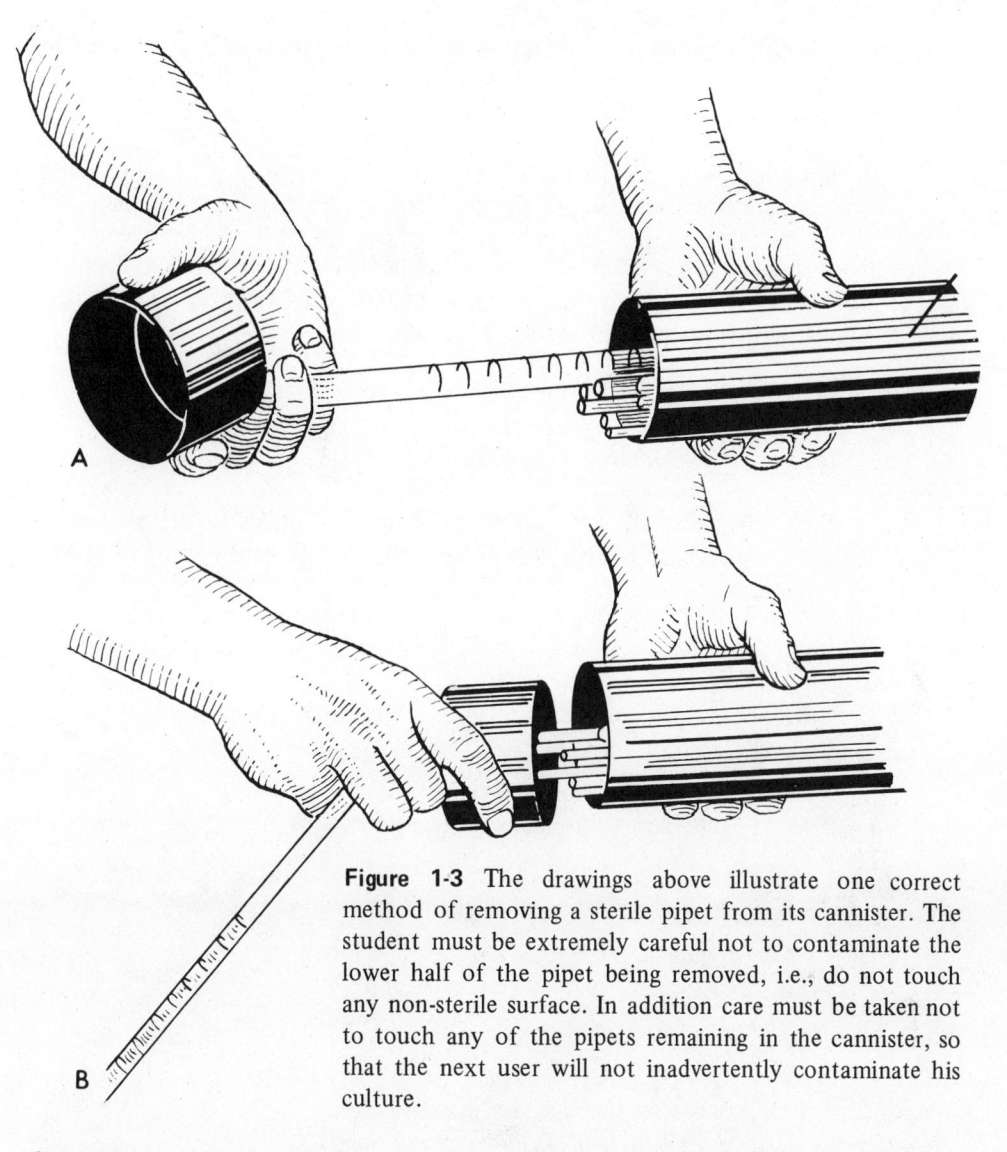

Figure 1-3 The drawings above illustrate one correct method of removing a sterile pipet from its cannister. The student must be extremely careful not to contaminate the lower half of the pipet being removed, i.e., do not touch any non-sterile surface. In addition care must be taken not to touch any of the pipets remaining in the cannister, so that the next user will not inadvertently contaminate his culture.

for. (b) The 10 ml. pipet is used for larger quantities and has as its principal calibration 1 ml. divisions, although it has 0.1 ml. calibrations as well. These last graduations are not as accurate, however, as those found on the 1 ml. pipet. In the next experiment you are given some practical uses for the bacteriological pipet. In addition, this experiment introduces the student to the concept of hydrogen ion concentration and the use of indicators and buffers.

In microbiology it is often necessary to measure the amount of hydrogen ion (H^+) or protons in a given solution, since all living forms are more or less sensitive to it. Theoretically this could be done by weighing the actual quantity of hydrogen ion, but in practice this procedure has many obstacles. A method has been devised making hydrogen ion measurement very convenient. A number system has been designed to indicate the actual quantity of hydrogen ion in a given solution in terms of *grams per liter*. Pure water has a hydrogen ion concentration of 0.0000001 gram per liter. This is a cumbersome number and it is more conveniently designated by the use of exponents — or simply 1×10^{-7} gram of hydrogen ion per liter. The use of the minus sign preceding the exponent indicates that the number is less than (1.0). Therefore, 1×10^{-7} actually means that the number 1 is seven places to the right of the decimal. In order to simplify this designation of quantity, chemists have created the so-called pH system, which makes a positive number out of the negative exponent. Thus one designates the pH of pure water as 7.0. As can be seen, one could trace this back ultimately to the long number mentioned a few sentences back. The pH designation, then, is a whole number that represents a very small quantity, and in reality is the positive number of the exponent. Note these examples:

(a) $0.000001 = 1 \times 10^{-6}$ = pH 6.0
(b) $0.0001 = 1 \times 10^{-4}$ = pH 4.0
(c) $0.1 = 1 \times 10^{-1}$ = pH 1.0

From these examples it is easy to see then that a pH of 1.0 actually means 0.1 gram of hydrogen ion per liter of solution. The pH scale is set up from 0 to 14 as follows:

```
0
1
2
3
4  ⎫ Acid (high hydrogen ion concentration,
5  ⎬       low hydroxyl ion concentration)
6  ⎭
6.9

7.0   Neutral

7.1 ⎫
8   ⎪
9   ⎪
10  ⎬ Basic (low hydrogen ion concentration,
11  ⎪       high hydroxyl ion concentration)
12  ⎪
13  ⎪
14  ⎭
```

It can be seen from this scale that solutions classified as *acids* contain hydrogen ion in quantities greater than 1×10^{-7} gram per liter, whereas solutions considered to be *bases* (alkaline solutions) contain less than 1×10^{-7} gram per liter of hydrogen ion. It is important to keep in mind that even though the pH numbers get larger, the *actual amount* of acidity of the solution gets *smaller*. Furthermore, it is important to recognize that the difference between a pH of 1 and a pH of 2 is not half as much hydrogen ion or twice as much, but rather *10 times* as little or 10 times as much. In order to see this clearly, express pH 1 and pH 2 as decimals. Therefore,

$$0.1 = \text{pH } 1.0$$
$$0.01 = \text{pH } 2.0$$

It can be seen then that the first number is 10 times more than the second number. Thus each number on the pH scale represents an interval of multiples of 10. Carrying this one step further, an interval of two pH numbers therefore represents 100 times as much and an interval of three pH numbers represents 1000 (10 × 10 × 10) times as much. In the final analysis, pH is a symbol designating the quantity of hydrogen ion in one liter of solution.

In order to be more complete on this subject, it should be mentioned that alkalinity in a solution is determined by the presence of hydroxyl ions (OH^-). A true understanding of this requires a knowledge of more advanced chemistry and is not extended here. For your own purposes, just recognize that a pH higher than 7.0 (i.e., 7.1, 8.0, 9.5, etc.) means that there are proportionately more OH^- ions in the solution than H^+ ions. The properties of an acid depend on the H^+ ion, and the properties of a base on the OH^- ion.

Indicators are substances widely used in microbiological work. They are colored compounds that ionize either as weak acids or as weak bases. The word "weak" means that the substances in question do not undergo ionization to a very great extent and therefore the quantity of H^+ or OH^- is very slight. All indicators change color, depending upon the pH of the solution in which they are dissolved. A typical indicator you probably are already familiar with is litmus. Litmus is used frequently to determine if a substance is acid or alkaline, according to the color the litmus assumes (red in acid solution and blue in basic solution). There are many indicators in use besides litmus. Each indicator exhibits a characteristic change of color and a characteristic pH range over which it shows this color change. One ordinarily speaks of a basic color and an acid color for each indicator. Some indicators display a distinct third color when the pH is midway between the acid and basic colors.

An indicator used quite frequently in microbiology is *phenol red*. Phenol red is yellow (acid color) in a solution with a pH less than 6.8, and it is red (alkaline color) in a solution with a pH more than 6.9. It actually assumes a magenta or violet-red color at extreme basic pH's. Phenol red is used in the next exercise.

Another concept with which the student should familiarize himself in the microbiology laboratory is that of buffer action. *Buffers* are salts of either weak acids or weak bases that tend to resist a rapid change in pH. Distilled water is an example of a non-buffered system, and the addition of a small amount of acid or base causes a marked change in the pH of distilled water. On the other hand, if one were to dissolve some sodium phosphate or some sodium bicarbonate in distilled water, he would find that now much more acid is needed to change the pH of the solution than was needed with the distilled water alone. Buffers are principally used in microbiological media to prevent a marked change in pH from the pH at which the medium was originally set. The end products of microorganism metabolism frequently cause acidic or basic changes that are detrimental to the further growth of the organism. If a buffering salt is first added to this medium, such a pH change is greatly slowed down or even eliminated.

There are many other substances that possess buffering ability. Proteins and amino acids are notable among this group. Proteins and amino acids tend to ionize either as acids or as bases in opposition to the pH of the solution in which they are dissolved. Therefore protein, if dissolved in an acid solution, ionizes as a base and tends to counteract the acidity of the original solution. The reverse is true if protein is dissolved in a basic solution. Since many of the ingredients of microbiological media are either protein or derivatives of protein (peptones, peptides, etc.), it is usually not necessary to add buffering salts to such solutions. However, when growing fastidious organisms, it is often desirable to add a buffering salt.

The following experiment is designed to introduce you to both the use of the bacteriological pipet and to the observation of indicators and buffers in action during alterations of pH in several solutions.

EXPERIMENT 1-2

The Use of Pipets, Indicators, and Buffers

MATERIALS

1. Distilled water.
2. 10 ml. pipets (4); 1 ml. pipets (2).
3. Solutions of 0.1 N HCl and 0.1 N NaOH.
4. Indicator solution of phenol red (0.02%) in dropper bottles.
5. Buffer solutions of 0.2 M K_2HPO_4 and 0.2 M KH_2PO_4.
6. Test tubes (4).
7. Peptone solution (8%) (protein-like).

PROCEDURE: In a test tube rack, set up four empty test tubes and number them from 1 to 4. Pipet 5 ml. distilled water into tube 1, to serve as a control. Into tubes 2 and 3, pipet 4 ml. of distilled water. Into tube 2, pipet 1 ml. of K_2HPO_4 buffer solution. Into tube 3, pipet 1 ml. of KH_2PO_4 buffer solution. Into tube 4, pipet 5 ml. of peptone solution. With a 1 ml. pipet add 10 drops of phenol red indicator solution to each of the four test tubes. Agitate the fluids so the contents of each tube are well mixed. Note and record the beginning color in each tube. Using a 1 ml. pipet, add 1 drop of the acid solution (HCl) to tube 1 and note the color after mixing. Do not use the same pipet for dispensing acid and base solutions. Now add 2 drops of the base solution (NaOH) to tube 1. Mix well and observe the change. Into tube 2, add acid solution one drop at a time until you get a color change. What does this color change signify? Make sure you mix the contents of the tube thoroughly after each drop. Count the number of drops required to bring about a color change. Into tube 3, add base solution one drop at a time, mixing the solution until a color change occurs. Count the drops. Explain what has occurred in each of the first three test tubes. Would you add acid or base to the peptone solution to test its buffering power? How many drops are required to change the indicator in this solution?

Bacteriological Incubators and Water Baths

Many microorganisms grow best at temperatures that approximate that of the human body (37°C. or 98.6°F.). To achieve and maintain this temperature, either the bacteriological incubator or a constant temperature water bath may be used. The incubator is very similar to an ordinary oven that is thermostatically controlled to maintain a constant temperature. Other microorganisms do not grow well at elevated temperatures and should be kept at 20° to 25°C. (68° to 78°F.). For such organisms, incubation should always be carried out at room temperature in your personal drawer.

For most purposes, the bacteriological incubator will serve your needs for elevated temperatures. Sometimes a more rapid transfer of heat is required than that provided by an incubator; in this case a 37°C. water bath is employed. One drawback of the bacteriological incubator is that it is a dry air device, which tends to dehydrate the various types of culture media. Therefore, any growth experiments utilizing the incubator should be *terminated* within 7 to 9 days at the longest. Be sure you do not leave any cultures in the incubator longer than this period, since they will dry up and harden on the glassware.

Bacteriological Sterilizers

It is necessary to completely eliminate all forms of life from the medium employed for growing microorganisms before it is used. Three methods of sterilization are mentioned here. For wet sterilization, the *autoclave* is used. The autoclave is a large pressure-cooker type device in which the boiling temperature of water is increased by an increase in pressure. The autoclave operates at 15 pounds per square inch (p.s.i.) above normal atmospheric pressure. At such a pressure, water boils at 121°C. (250°F.) instead of 100°C. (212°F.) as at normal atmospheric pressure. All forms of life are usually killed by maintaining a temperature of 121°C. for 15 min. This is the temperature-time cycle found in the bacteriological autoclave.

Another method of sterilization is *dry sterilization,* for which the common household oven may be used. Glassware and other inert objects are sterilized in this fashion. To sterilize with dry heat, high temperatures must be maintained for longer periods as compared to autoclaving. The typical sterilizing cycle in the dry-air oven is 180°C. (350°F.) for 3 hours.

Solutions that are destroyed or decomposed by autoclave temperatures can be sterilized by passing them through bacteria-tight filters that prevent passage of all bacteria and larger microorganisms. Fluid passed through the filter is quite sterile and need not be treated by heat or chemical action. Many laboratories use both the Seitz filter (made of asbestos) and the cellulose membrane filter.

The Inoculating Loop and Needle

Inoculating loops and needles are often used for the purpose of transferring microbial growth from one container to another without introducing unwanted organisms (contaminants). Each of these devices consists of a handle to which is attached a piece of wire made of some inert metal such as nichrome or platinum. The wire may be either straight (needle) or looped at the end. Each type serves a distinct purpose.

The wire is always sterilized in a Bunsen burner by flaming to red heat. The wire is then allowed to cool for about 10 to 15 seconds. When using the Bunsen burner, adjust the flame so you have a distinct, blue cone; then pass the wire to be sterilized through the tip of the blue cone to obtain rapid heating. The wire, now sterile, completely free of any living organisms, can be used to fish out bacterial growth and transfer it aseptically to another container.

Bacteriological Slides

During the semester, various types of slides will be used. The most common type is the plain glass slide. Another slide frequently used is the *concave* slide – a clear glass slide with a hollowed-out portion in the center. This will be used in later experiments for the observation of living bacteria. A *deep well* slide may also be included in your slide box. It is used for growing a microculture of fungi. It is not to be confused with the concave slide. The deep well slide has the entire upper surface frosted and the depression is cylindrical rather than concave.

INTRODUCTION TO SOME MICROBIOLOGICAL TECHNIQUES

Probably the most frequently used technique in the microbiology laboratory involves the transfer of microbial growth from one environment to another. There are many microorganisms in the air around us and on all surfaces in our environment. Because of this, it is necessary to set up adequate precautions to prevent unwanted organisms from getting into experimental tubes and plates. Once the appropriate steps have been taken to isolate and propagate a pure strain of microorganism, the student must be careful that contaminant organisms do not get mixed in with them. It would be simple enough just to leave the pure strain in the original container, but microbes, like people, must have both a continual supply of food or nutriment and a means of ridding themselves of their own waste products. After a population of microorganisms has grown for a certain length of time in the same culture tube, it must be transferred to a fresh food supply if the culture is to survive. For this, as well as other reasons, several techniques for inoculation and subculture have been developed.

Although use of the inoculating loop and needle is perhaps better demonstrated than described, a brief explanation is in order. The inoculating

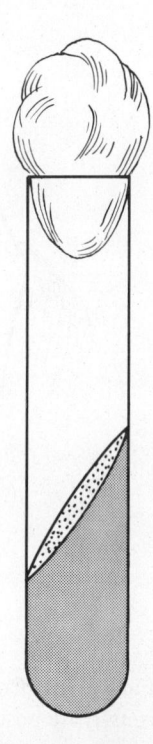

Figure 1-4 An agar slant showing the growth of bacteria on the surface of the agar. No organisms are growing *in* the agar, thus the inoculating loop used to retrieve growth should not gouge beneath the surface.

loop is ordinarily used to transfer a culture from one test tube to another. The growth of microorganisms on an agar slant is "heaped up" on the surface of the medium. *None* of the organisms is buried in the agar and therefore, when removing growth from an agar slant, the student should touch *only* the material on the *surface of the agar*. There is a tendency among beginning students to dig down into the agar with the inoculating loop in order to obtain a sample. This practice only makes a mess out of the agar slant and does not accomplish the original intention.

A common misconception of the beginning student is that a large amount of growth must be fished out to make a successful transfer. Bacterial growth equal to the size of a pinhead probably contains several million organisms. It is therefore obvious that very small (barely visible) quantities adhering to the inoculating loop are sufficient to effect a successful transfer.

Step by step, the transfer procedure begins by flaming the inoculating loop to red heat and allowing a 10 to 15 second cooling-off period. The culture tube from which the organisms are to be obtained is then picked up by the other hand, and the cap (or cotton plug) is removed. Keep the cap in the crook of the little finger of the right hand (for a right-handed person) until time to replace it on the tube. Never lay a cap or plug on the table, because this would surely contaminate the agar or culture. The neck of the tube is passed back and forth through the flame of the Bunsen burner once or twice slowly to destroy any unwanted organisms adhering to the top of the tube. The sterilized loop is then slowly inserted into the test tube and a small amount of microbial growth obtained with the loop. Upon removal of the loop, the tube is flamed again and the cap replaced. Then the new tube is picked up, the cap removed, and the neck flamed as with the first tube. The

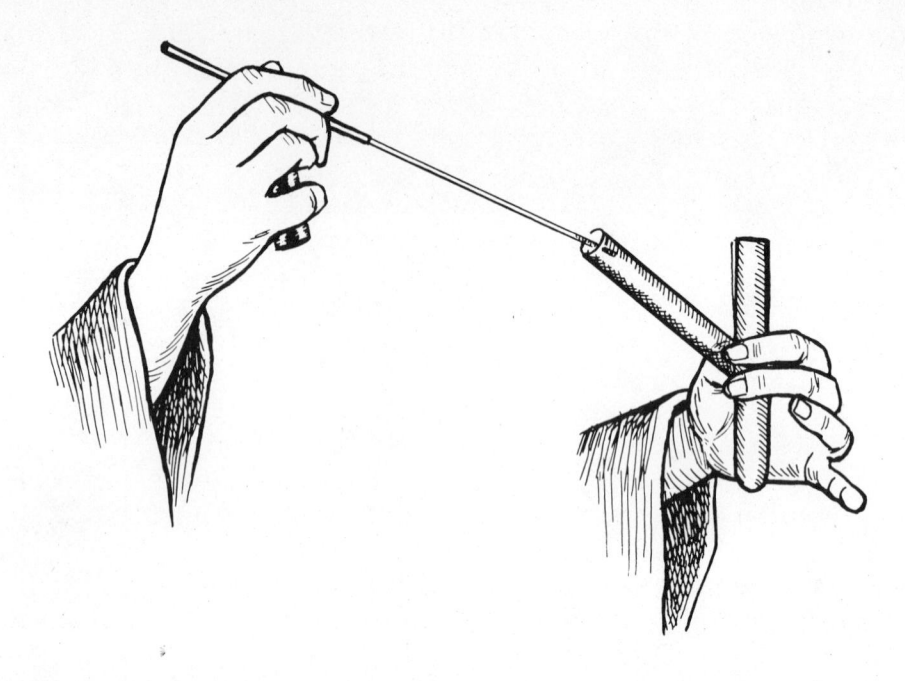

Figure 1-5 Correct method for holding inoculating loop during transfer of microorganisms. Note position of the two test tubes and the test tube closure.

inoculating loop is inserted into the new tube and the organisms deposited *into* the medium if it is a broth or a fluid culture, or streaked *gently* back and forth on the surface if it is an agar slant culture. The loop is removed, the neck of the tube flamed, and the cap replaced. Finally the loop is flamed to red heat to destroy all organisms adhering to it. This is the basic technique that, if properly performed, will permit you to transfer safely more than 99 per cent of all microbial cultures. Other methods of inoculation will be demonstrated by the instructor as needed.

The Isolation of a Pure Culture of Bacteria

Any organized study of living forms must be planned so one species at a time is examined rather than a mixture of many different kinds. One of the most frequent problems in microbiology is the isolation of a pure culture of a given bacterial species. Ideally, one would remove a single bacterium to a sterile environment until it divided to form many more bacteria, all derived from the original one. Such a population is called a *clone* and one would know that the heredity of all the members of this population was identical, barring mutations. But it is very difficult to isolate a single cell, although there are techniques available that approximate this even if single-cell isolation can't be proved.

Undoubtedly the most widely used technique for the isolation of a pure culture is the so-called *streak plate* method. Ordinarily a Petri dish is employed because of its broad surface. A *small* amount of growth is obtained on a sterilized inoculating loop, and the loop is then dragged lightly over as much of the surface of the agar in the Petri dish as possible, with the hope that the bacteria dropping off of the loop by this action will finally begin to drop off *one at a time*. After suitable incubation, these single deposited cells divide many times and finally form a visible mass of growth on the agar surface originally derived from the (presumed) single organism. This visible mass of growth on the agar is called a *colony*. Growth from this colony (pure culture) is then transferred into a separate, sterile medium as a future source of this organism (a *stock culture*). This and other techniques involving the use of the Petri dishes are covered in the next experiments.

EXPERIMENT 1-3

Isolation Techniques and Use of Petri Dish Cultures

MATERIALS

1. Sterile Petri plates (4).
2. Nutrient agar* (bottle) and nutrient agar slants (2).
3. Bent glass rod.
4. Sterile cotton tip applicator stick (1).
5. A mixed broth culture of Aerobacter aerogenes** and Serratia marcescens.***

Streak Plate Method of Isolation

PROCEDURE: Melt nutrient agar in the bottle. This is done easily by putting the bottle with cap loosened slightly in a can of water and placing the can on a tripod stand over a Bunsen burner. A supporting material such as wire gauze is often used on the tripod. Make sure the water level in the can is at least as high as the agar in the bottle. A bottle containing about 150 to 200 ml. of solidified agar usually takes about 20 to 30 min. to melt in boiling water, at which time the agar appears quite clear and translucent. Remove the bottle from the water and allow it to cool to approximately 45°C. This temperature

---

*Traditionally, courses in microbiology have used a standard nutrient preparation for routine use that goes by the name Nutrient Agar. In truth this preparation is not very "nutrient" for a large number of organisms. There are a number of formulations available which are far more satisfactory for routine use, such as Trypticase Soy Agar (BBL) or Tryptic Soy Agar (Difco). Alternatively, Nutrient Agar (or Broth) can be augmented by the addition of 15 gm. of a peptone, 5 gms. of yeast extract and 2 gms. of glucose per liter. Any future reference to nutrient agar or broth in this text will be made purely for purpose of avoiding nomenclature difficulties, and with the understanding that one of the more suitable preparations mentioned should be employed.

**Aerobacter aerogenes* is a small rod-shaped bacterium that is widely distributed in nature. It is often associated with the intestinal canal of man and various animals and is found on plants and in waters, milk and dairy products.

***Serratia marcescens* is a small rod-shaped bacterium that is widely distributed in nature in much the same places as *Aerobacter aerogenes*. It is well-known in the bacteriologic laboratory because it is frequently an airborne contaminant and produces a red-pink pigmented colony. The pigment is named prodigiosin.

can roughly be determined by holding the bottle firmly in your hand. If it is slightly warm to the touch but not hot, it should be just about right. Do not allow it to get too cool. Remove the cap from the bottle of agar, flame the neck of the bottle and pour about 15 to 20 ml. of the melted agar into a sterile Petri dish. Immediately cover the dish and allow it to sit on the table top for about 15 to 20 min. so the agar will harden properly. Flame and cap the agar bottle. Remove a small amount of growth with your sterilized inoculating loop from the mixed culture provided and trace a surface design on the agar according to instructions.

Several patterns can be used, some of which are illustrated in the diagrams below.

Keep the lid of the Petri dish in a slanted position just above the agar so as to protect the agar surface during each phase of the streak-plate operation. The student must be careful to avoid digging or gouging the agar during streaking, because this removes most of the organisms at that point. The general idea in most streak plate methods is to remove as much growth as possible in the first section (1) of the streak plate and then to begin tracings that do not recross any previous tracing. Your best isolation should come at the end of the streak (section 4).

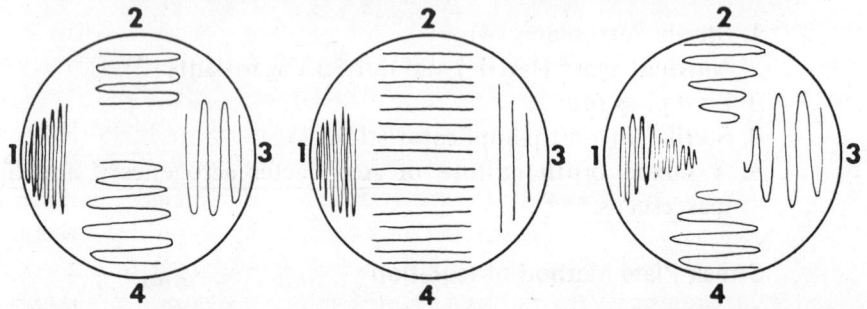

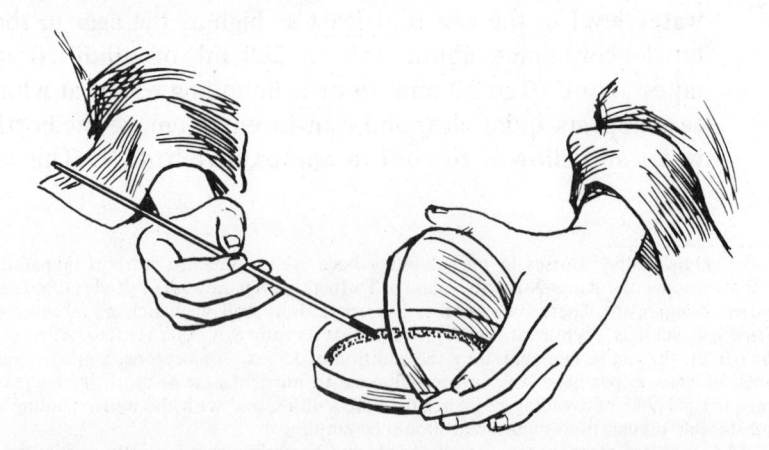

Figure 1-6 Correct technique for streaking a Petri plate of nutrient agar for isolated colonies. Note the comfortable position of the right hand and the protective angle of the lid held by the left hand. When one sector is completed, the loop is withdrawn and the lid closed. The plate is rotated 90° and the next sector streaked.

The colonies formed from the two organisms in the original broth culture will be colored differently when growth is complete, making it easy to determine if you have done a proper job of separating them. Label the streak plate with your name and the number of the experiment, incubate at room temperature in an <u>inverted</u> position, and allow it to grow for 24 to 48 hours. Petri dishes are usually incubated in the inverted position to prevent water condensation (which forms on the under side of the lid when right side up) from dropping onto the agar surface. The condensation would mix all organisms on the plate and ruin any attempt at isolation of pure colonies.

Observe the results and look for well isolated colonies of both kinds of bacteria. Because this experiment is primarily a procedural one, no formal laboratory report is necessary. However, do not be satisfied with a poor result. If isolation of a reasonable number of colonies is not obtained, repeat the experiment until you have mastered the technique. The instructor will check your plate to determine if you have succeeded. In many of the later experiments the entire burden of success or failure rests upon <u>first</u> isolating an organism from a mixture of bacteria.

The student can ordinarily determine when a colony has grown from a single bacterium of either of these organisms; the pure colony tends to be very regular (round) in shape and does not touch any other colony on the plate. Any colony that is in direct contact with another colony is <u>not suitable</u> and cannot be used. Select one colony of each type that is well isolated and transfer a little material from each colony with the inoculating <u>needle</u> to separate slants of nutrient agar, using the usual aseptic precautions to prevent contamination. Streak the surface of the agar slant in a zig-zag fashion to get a maximum amount of growth. Isolation is not the purpose here,

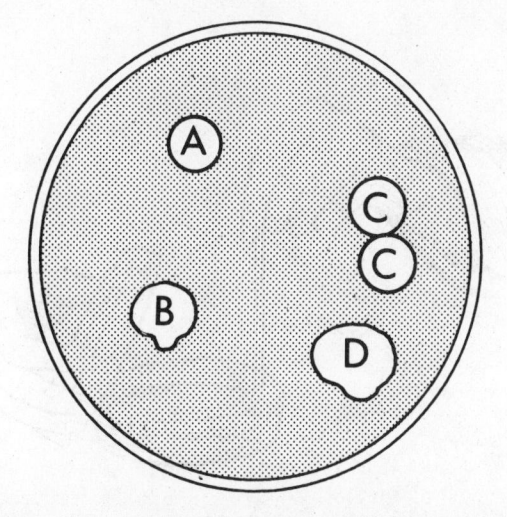

Figure 1-7 Examples of colonies which should not be picked for subculture (B, C and D) and one that is satisfactory (A).

simply maximum growth. Now incubate for 24 hours at room temperature.

The growth in one tube should appear entirely white and the growth in the other tube should appear entirely red or pink. Each culture would now be designated as a stock culture of the organism in question, since you could now remove a sample of growth from either tube and be assured that you have a pure culture of that particular organism. This procedure is utilized many times during the semester to obtain a pure culture. Observe the growth on each slant and then discard the media in the tubes and the Petri plate according to the method prescribed by your instructor.

Isolation of Bacteria with a Bent Glass Rod

PROCEDURE: Another method of isolating microorganisms and growing pure colonies involves the use of the bent glass rod. Prepare a Petri dish for inoculation in the usual fashion. Sterilize your inoculating loop and deposit a single loopful of the mixed culture provided near one edge of the agar in the Petri dish. Pour one or two drops of 70% alcohol from the reagent shelf on the short, bent part of the glass rod. Light the alcohol on the rod by touching it to a Bunsen burner flame. Permit it to burn completely. Repeat this burning procedure. Permit the glass to cool in the air for about one minute. Then, spread out the culture deposited on the agar in the Petri dish over the entire surface with the short arm of the bent glass rod. Ordinarily this should distribute single organisms into various portions of the agar sufficient to give isolated colonies. <u>Invert the dish</u>, mark it appropriately, and incubate for 24 to 48 hours at room temperature. Observe for isolation as you did in the streak plate experiment. Your instructor will check your plate for isolation.

Figure 1-8 Correct handling of the bent glass rod for spreading microbial growth on the surface of a Petri plate of agar.

Sectoral or Zone Growth

PROCEDURE: The Petri dish is very useful in addition to isolating pure cultures. The broad surface of the dish enables one to perform multiple experiments on it if properly used. As a demonstration of this, prepare a Petri dish with nutrient agar as done previously. After the agar has hardened, invert the plate and draw lines on the bottom of the plate so that it is divided into 6 equal sectors as if you were cutting a pie. Number the sectors from 1 to 6. Turn the plate over and streak sector 1 with a drop of tap water, tracing a zig-zag line within the confines of the zone. Place a drop of water on the skin of your forearm, rub it several times with the loop, remove a sample of this water, and streak it on sector 2. In similar fashion get a sample of water from the floor for sector 3 and from a human hair for sector 4 (immerse a hair in a drop of tap water on a slide and then streak). On sector 5, streak a sample of saliva; leave sector 6 blank. Invert the plate and permit it to incubate in the 37°C. incubator for 24 to 48 hours. (Unless otherwise specified, incubate all future Petri dishes inverted.) Observe the various types of microbial growth obtained in this fashion. Note that there is little, if any, isolation. This method does, however, permit multiple inoculation on a single piece of glassware. Later experiments utilize the sectoral or zone method for this and other types of growth experiments. Again, no formal laboratory report is necessary here.

Isolation of Microbial Growth from a Cotton Swab

PROCEDURE: A technique commonly used in medicine is the procuring of microbial specimens on a cotton swab. Since streaking the surface of a Petri plate with such a device would only give a grand mixture of organisms with no isolated colonies, this is not a desirable method. There is, however, a technique that enables isolation of pure colonies from a cotton swab. Procure a sterile cotton swab and immerse only the tip of it into the mixed broth culture provided. With a back-and-forth motion, streak a _small_ sector on one edge of a Petri dish prepared with nutrient agar. Close the Petri dish and throw the swab into the wastebasket. Sterilize the inoculating loop and drag it through the sector streaked with the cotton swab to pick up organisms deposited there; then streak into the uninoculated area in a fashion similar to that performed in the streak-plate technique. Be sure that you do not re-introduce the loop into the swabbed area after the first two or three passes through it. If this is performed properly, isolated colonies should appear in the usual fashion. Incubate inverted for 24 to 48 hours at room temperature and examine for isolated colonies. No report is needed for this experiment, but your instructor will check your plate.

Sterile Technique

The problem of sterile technique has already been introduced as one that constantly faces the microbiologist in the laboratory. Despite the need for careful attention to detail in this respect, it is often difficult for the beginning student to realize just how important attention to detail really is. This is probably because the student is not sufficiently aware of the ubiquity of microorganisms. Every surface, every drop of fluid, every portion of air — all are teeming with microbial growth. Only after the student realizes that he is living in an environment that is liberally populated with microorganisms, can he fully appreciate the necessity for aseptic or sterile technique. No amount of talking or reading about this problem impresses the beginner nearly as much as first-hand experience. Since these organisms are invisible to the naked eye, experiments to demonstrate their presence are based on showing results of their growth. In Experiment 1-4, you will see evidence that indicates the presence of these organisms in and around us, as well as the necessity for adhering very strictly to rules of sterile technique in the microbiological laboratory.

EXPERIMENT 1-4

The Omnipresence of Microorganisms in the Environment and the Necessity for Aseptic Technique

MATERIALS

1. Petri dishes, sterile (1) and nonsterile (1).
2. Sterile nutrient agar for plating.
3. Tubes of sterile nutrient broth (2).
4. Sterile pipets (1.0 ml.) (1).

PROCEDURE: Wash the nonsterile Petri dish until it is thoroughly clean. Melt some nutrient agar in your bottle. After the agar has cooled to the proper temperature, pour 10 to 15 ml. of it into the Petri dish, immediately close the bottle, and permit the agar in the Petri dish to harden. Label this dish "clean but not sterile." Into the sterile Petri dish pour another 10 to 15 ml. of agar. After this has hardened, open the dish and leave it exposed to the air for 30 min. Select a spot different from those selected by other students in the class so that all plates are not exposed to the same environment. Suggested places include: the edge of the hallway floor, the laboratory floor, and outside the building. Label this plate "Sterile but open to air."

Using the two sterile test tubes of nutrient broth, transfer 1 ml. of broth with a 1 ml. pipet from one tube into the other tube without observing the usual precautions — that is, do not flame the neck of the tube, etc. Next, transfer 2 ml. from the second tube back into the first in the same fashion using the same pipet. Repeat this basic maneuver five more times without utilizing aseptic technique. Do not hurry. Now cap the tubes and put them in your drawer for 48 hours along with the Petri plates you have prepared. Sometimes it is even desirable to let these tubes and plates incubate for at least 4 days. After incubation, examine the plates and tubes for evidence of growth, and record your observations on each culture together with your interpretations of the results. In addition to the observations on the presence of microorganisms in your environment, compare the various colony forms that have developed in the Petri dish you exposed to the air with the diagrams that appear on page 26, entitled, "Cultural Characteristics of Bacteria." Select three distinctly different colonies from the agar plate and characterize each colony completely, using the descriptive terms given on the diagrams. Each colony can be studied in terms of its overall form, its elevation, and its margin. In order to do this, you should use the low power objective of your microscope, as well as the naked eye. You will have to remove the top of the Petri dish and the mechanical stage to make room for the Petri plate on the microscope stage. Be careful not to lower the objective lens into the agar. In addition to providing colony studies, the diagrams also help categorize bacteria according to the way they grow under other environmental conditions. An agar stroke, for example, is prepared by simply dragging inoculum in a straight line with the wire loop from the bottom of the nutrient agar slant to the top. The type of growth attained after incubation is then observed. The other diagrams are referred to in later experiments.

Anaerobic Culture Methods

Most living organisms, other than plants (and even they need oxygen in the dark), require a constant supply of oxygen to survive. However, this is not always the case with microorganisms. Some microorganisms not only do not require oxygen but are actually poisoned by its presence. Such organisms are said to be *anaerobic* ("without air"). In general, microorganisms can be classified into three categories according to their oxygen requirements. Microorganisms that require air are called *aerobes;* those that cannot live in the presence of air are called *anaerobes;* and those that can live either with or without air are said to be *facultative.* Members of a fourth, occasionally used classification are said to be *microaerophilic.* Such organisms ordinarily prefer an anaerobic existence, although they generally can survive on a restricted basis in the presence of air. In the subsequent experiment you will observe some of the different ways in which anaerobic organisms are cultivated.

CULTURAL CHARACTERISTICS OF BACTERIA

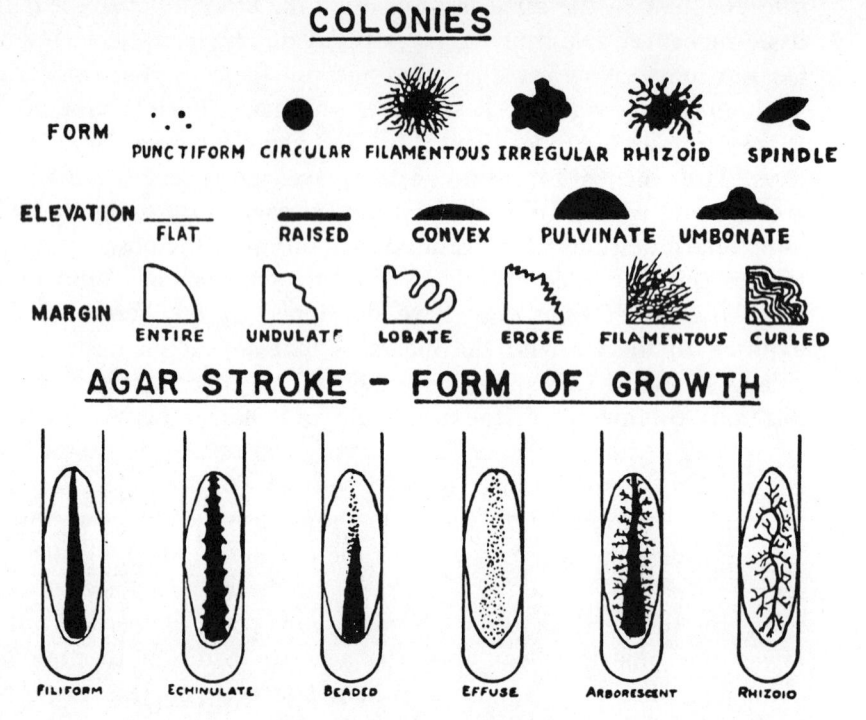

COLONIES

FORM: PUNCTIFORM, CIRCULAR, FILAMENTOUS, IRREGULAR, RHIZOID, SPINDLE

ELEVATION: FLAT, RAISED, CONVEX, PULVINATE, UMBONATE

MARGIN: ENTIRE, UNDULATE, LOBATE, EROSE, FILAMENTOUS, CURLED

AGAR STROKE – FORM OF GROWTH

FILIFORM, ECHINULATE, BEADED, EFFUSE, ARBORESCENT, RHIZOID

GELATIN STAB

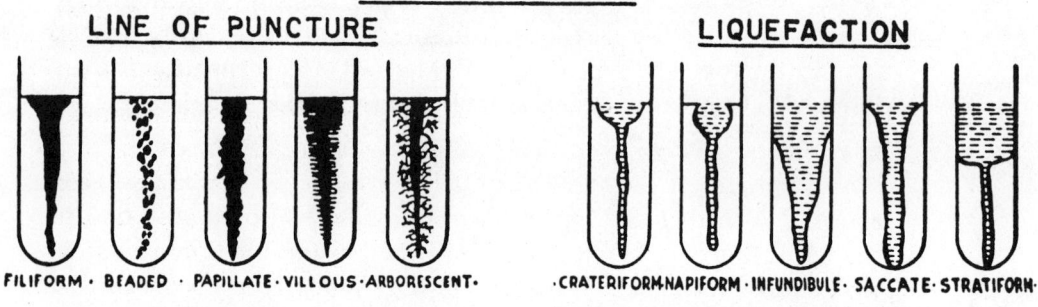

LINE OF PUNCTURE
· FILIFORM · BEADED · PAPILLATE · VILLOUS · ARBORESCENT ·

LIQUEFACTION
· CRATERIFORM · NAPIFORM · INFUNDIBULE · SACCATE · STRATIFORM ·

NUTRIENT BROTH – SURFACE GROWTH

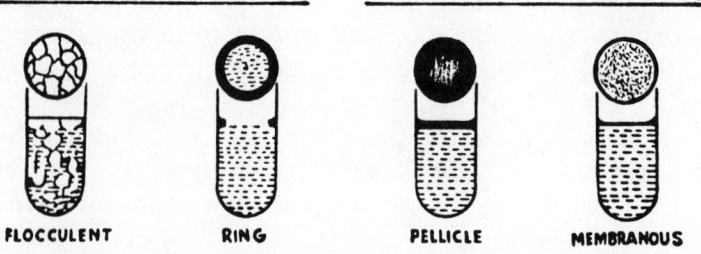

FLOCCULENT, RING, PELLICLE, MEMBRANOUS

Reprinted from Leaflet V of Manual of Methods for Pure Culture Study of Bacteria

(The above chart provided through the courtesy of the American Society for Microbiology, Detroit, Michigan.)

EXPERIMENT 1-5

The Cultivation of Anaerobic Organisms

MATERIALS

1. Nutrient agar slants, cotton stoppered (2).
2. Nutrient agar talls (2).
3. Thioglycollate broth tubes (2).
4. Pyrogallic acid (powdered).
5. Sodium hydroxide solution (4%).
6. Medicine droppers.
7. Rubber stoppers (6).
8. Cultures of Clostridium sporogenes* and Bacillus subtilis.**

PROCEDURE: Inoculate one tube of thioglycollate broth with Clostridium sporogenes and another tube of broth with Bacillus subtilis. Be careful not to agitate these tubes once they have been inoculated. Using the inoculating needle, inoculate one agar tall with Cl. sporogenes and another with B. subtilis. Stab down deep into the center of the agar, then withdraw the needle along the line of the inoculation. Label each tube with the name of the organism contained and put the tubes in the 37°C. incubator. Streak one sterile nutrient agar slant with Cl. sporogenes and another with B. subtilis. (In this particular experiment cotton stoppers must be used in these tubes.) Push the cotton stopper into each tube with the handle of your inoculating loop until the cotton almost touches the agar in the tube. Above the cotton stopper add dry pyrogallic acid, making a layer about half an inch deep. Allow room for the insertion of a rubber stopper at the top of the tube above the pyrogallic acid. Add about two medicine droppers full of sodium hydroxide solution to the pyrogallic acid in each tube, and stopper the tubes immediately with rubber stoppers and invert. Put both properly labelled, inverted tubes in a can and place both tubes and can in the 37°C. incubator for 24 to 48 hours. The pyrogallic acid-sodium hydroxide combination is an efficient oxygen absorbent. Therefore any growth in either of the tubes just inoculated must take place without any oxygen. Sketch in your laboratory notebook the growth patterns you obtain in all six tubes and briefly state your conclusions.

Clostridium sporogenes is a spore-forming rod which is an obligate anaerobe. It is a very common soil microorganism, particularly one that is well fertilized. It will characteristically produce foul odors when decomposing various kinds of organic matter.

**Bacillus subtilis* is a widely distributed soil form and is also a very common air contaminant. It is a fairly large spore-forming rod.

CHAPTER 2

CYTOLOGY OF MICROORGANISMS

The word *cytology* means "the study of cells." Cytology is similar to the subject called gross anatomy, but on the microscopic level. Cytology is concerned with the cell and its parts; however, the structures studied in cytology are quite different from those in anatomy, for they are microscopic and are not nearly so clearcut. In the development of the science of cytology, scientists have had to resort to a great number of chemical "tricks" in order to visualize the structures within the cell. Until recently, cytology has been greatly hampered by a general lack of knowledge of the chemical nature of subcellular structure. Recent work emerging from biochemistry, however, has clarified many of these problems.

One of the principal tools in the study of the cell is differential staining. A *differential stain* is one in which two or more reagents are applied to impart color to a cell or its parts, whereas a *simple stain* is the application of a single stain or reagent. You may know that when you wish to detect the presence of starch in a structure such as a potato, you add an iodine solution to it. If starch is present, it forms a very dark blue or black colored complex. This test is considered to be specific for starch, because iodine gives this color only when it comes into contact with starch. Therefore, a beginning lesson in staining might be the application of an iodine solution to sample cells to detect starch particles within the cell. The application of an iodine solution to a cell normally gives the cell a brownish color characteristic of the iodine solution, but any starch granules in the cell show up dark blue or black, depending upon the strength of the iodine solution.

The application of certain chemicals to different substances produces color reactions specific for the treated substances, as in the case of iodine and starch. A great number of specific tests are now known that stain certain compounds or types of compounds. These tests have proved invaluable in cytology. With the gradual accumulation of large numbers of such tests, it has become possible to identify the chemical constituents of the cell, such as lipids, nucleic acids, and others. It has also been known for some time that cellular proteins and amino acids react as weak acids or bases with a great variety of materials. In other words, proteins are *amphoteric* substances; they can ionize either as an acid or as a base. This means that proteins are highly reactive chemicals.

> Ionization is the process by which atoms or molecules acquire an electric charge. Such charged particles are designated as follows: Na^+, H^+, OH^-, Ca^{++}, Al^{+++}, etc. Note that the *degree* of charge is indicated by the number of plus or minus symbols. The type of charge is also indicated. Ionization is the result of the loss or gain of electrons, thus changing the electrical balance of the particle in question. An *ion* is defined as an electrically charged atom or molecule.

Since proteins constitute the bulk of the solids in the cell, they are involved in most of the staining reactions. In order to react successfully with a protein, a biological stain must display a color and be able to ionize either as an acid or as a base. These substances are generally called *basic dyes* if they have a strong affinity for acid portions of the cell and *acid dyes* if they have a strong affinity for the basic portions of the cell. Many times it is useful to use a mixture of two dyes possessing different colors, one being acid and the other basic. When this type of stain is used, a cell with an acid part and a basic part appears with these respective parts stained in different colors. The well known hematoxylin-eosin stain used in the staining of animal tissue is a good illustration in point. Hematoxylin is a basic dye and imparts a blue color to the acid portions of a cell; eosin is an acid dye and colors the more basic areas pink or red. Since the nucleus of a cell is acid (due mainly to deoxyribonucleic acid, or DNA) and the cytoplasm more basic, simultaneous application of these two dyes to an animal cell generally shows a blue nucleus and a pale pink cytoplasm, although variations from this pattern are encountered.

In the application of these techniques to the study of microorganisms, the student is confronted with special problems. One outstanding problem is the minute size involved. Microorganisms are considerably smaller than either plant or animal cells, and differentiation of a structure within such tiny cells becomes very difficult with the usual magnifications given by the optical microscope. Another problem arises from the fact that microorganisms are not as clearly differentiated within the cell as are cells of higher forms. Thus discrete observation of internal structures is difficult. However, great progress has been made through an ingenious combining of techniques to visualize many structures in the cells of microorganisms. This progress has been greatly augmented by electron microscopy, but this technique is beyond the reach of most student laboratories.

Certain techniques and procedures are common to all cytological studies. One of these techniques involves the preparation of cells for observation. A problem that must be solved in the preparation of a microscope slide is the deposition of a sufficiently thin layer of cells so light can pass through, making them visible through the microscope. In plant and animal tissue, the usual procedure is *thin sectioning* with a *microtome,* a technique in which the tissue is sliced into very thin layers and then stained before observation with the microscope. Such sections are usually 5 to 10 μ thick. When dealing with microorganisms, this problem is considerably simpler. Here one needs only to perform a *smear preparation* for most stains. Since the cells are ordinarily quite small, deposition of a thin layer of them in a liquid suspension on a slide and allowing them to dry make it possible to observe these organisms without the aid of specialized equipment. It is possible to make a smear preparation from either plant or animal tissue, providing some means of dislodging the cells is found.

When a smear preparation is made from a broth culture of microorganisms, the broth suspension is applied directly to the slide *without prior dilution.* It is often helpful to use 2 to 3 loopfuls on a small area of the slide. The suspension should not be spread out too thinly. When making a smear from *solid media,* the problem is in getting *too many* cells. Only a very small portion of growth (about the size of the head of a pin) should be removed. The portion of growth substance should be emulsified in a drop of tap water on the slide, then spread out over a fairly large area. When dry, it should look *faintly* white, like a hard-water smudge spot or dilute milk.

Once a smear preparation is completed, it must be made certain that the cells will stuck to the slide and not be washed off during staining procedures. To do this, the student will use the technique known as *fixation*. Fixation is the process that stops all enzymatic activity in the cell before observation. Most, if not all, cells contain proteins called enzymes that are capable of destroying much of the material within the cell after it dies. Therefore, if a cell is not properly fixed before staining, much detail of cell structure will often be destroyed by these enzymes. In addition, heating helps to make the cells adherent to the slide and thus minimizes their being washed off during staining.

In microbiology, this cell structure destruction is avoided by use of heat fixation, whereas in plant and animal tissue, it is ordinarily done by the use of special chemicals. Most chemical fixatives are well known. They include formaldehyde, alcohol, and other substances of this type. Perhaps it would now be in order to undertake a few simple experiments that illustrate some of the ideas that have been discussed here.

EXPERIMENT 2-1

Smear Preparations and Fixation

MATERIALS

1. Alcohol (95%).
2. Microbial culture on agar slant.
3. Applicator stick.
4. Loeffler's methylene blue stain.
5. Toothpick.

PROCEDURE: Thoroughly clean three glass slides with soap and water, followed by the application of several drops of 95% alcohol

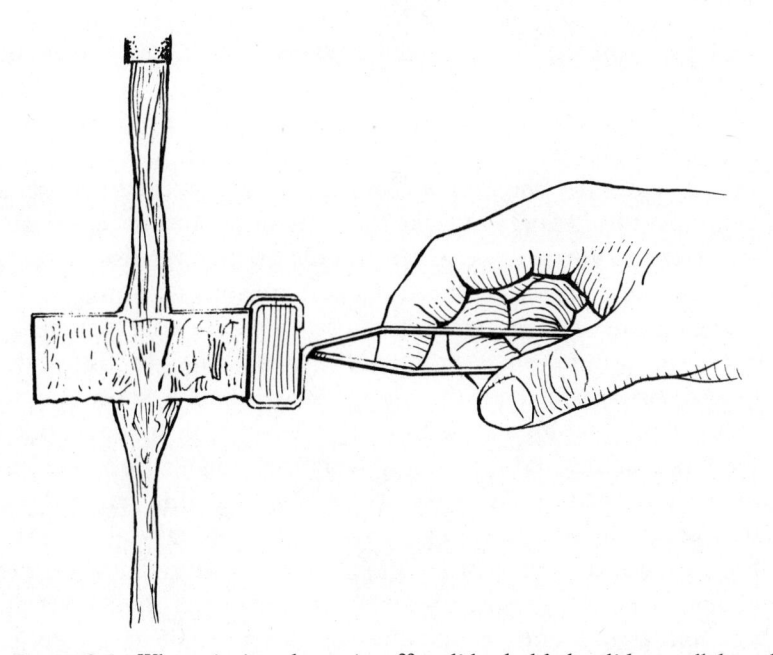

Figure 2-1 When rinsing the stain off a slide, hold the slide parallel to the flow of water as shown above. This reduces the loss of organisms from the preparation as compared to letting the water hit the smear directly. This method of rinsing is particularly necessary when the number of organisms is minimal, i.e., when taken from a broth culture.

from the reagent bottle. Allow the slides to dry. Then flame the "up" side of each slide for a moment in your Bunsen burner and allow to cool in the slide holder. You should use this method of slide cleaning in all future experiments unless otherwise instructed. A dirty slide often causes a considerable loss of material and makes it difficult to locate specimens for observation. On slide 1 place a single drop of tap water using the inoculating loop. Emulsify with the water a small quantity of the microbial growth provided and allow to dry in the air. On slide 2 repeat the procedure, except this time emulsify material obtained from between your teeth using the applicator stick provided. Next, gently scrape the inner lining of your cheek with a toothpick and deposit the material in a drop of water on slide 3. In each case spread the material out thinly so it is not concentrated in a single spot. After all three slides have dried, fix the first two slides with heat. This is done simply by grasping the slide with your slide holder and passing it back and forth through the flame of your Bunsen burner rather slowly three times. The glass should not be permitted to get too hot to touch. On slide 3, apply a drop or two of alcohol for 1 min. Allow to dry. Now apply a drop or two of the Loeffler's methylene blue stain to each slide and allow to react for 5 min. Gently rinse off the stain with tap water by holding the slide parallel to the stream of water briefly and allow each smear preparation to dry. Examine your slides using first the low power objective, then high power, and finally oil-immersion. Note the differences in size of the various types of cells under each objective. Make drawings of your observations under the oil-immersion objective. Observe the size and structural differences between cells of the different types.

As you may note, microorganisms are exceedingly small and detail is quite difficult to see, let alone draw. In this and in all future experiments you will be asked to enlarge the organisms over the size that you actually see so that the instructor can determine if you have observed the various structures correctly. Be sure you make your drawings definite in outline, not just sketches. In this and in any future observations of microorganisms when a stain is employed, apply a fill-in color to the drawing of the organism corresponding to the color of the stain. In many instances you will observe hundreds of bacteria in the microscopic field. The question naturally arises as to whether or not you must draw all of them. Obviously, this would be impractical. You will be asked to select several typical examples of the organisms you are observing, looking carefully for variations in size and shape, and also paying particular attention to any possible characteristic groupings or arrangements the organisms may assume on the slide. Many microorganisms are partially characterized by the types of arrangements they show in such preparations; therefore this is an important part of microscopic study of these organisms. Another problem that arises frequently is the over-concentration of organisms in the center areas of a stained preparation of bacteria, resulting in improper observation. In this case the student should search for an area where the organisms are distinct and individual. This is usually best found near the edge of the preparation where the cell population is considerably lower. Do not attempt to make any drawings of an amorphous mass of color.

NOTE: The following procedure should be used for the disposal of used slides. Place several drops of xylene on the slide, holding it over the sink. Two or three drops should be sufficient. Rub off the smear that you have prepared, using soap and water. Repeat this at least once to insure complete removal of all organisms. Then clean the slide as usual in preparation for re-use in other exercises.

Sometimes in cytological studies, it is important to obtain a more accurate picture as to the true size of the cell. This can be accomplished in several ways. One method employed frequently in microbiology is that of the negative stain. In this case the organisms are *not* stained, but the glass on which they are deposited is stained. In this way the student can observe the shape or outline of the organism clearly as a "window" in a dark background. The size of the organism is much more accurately determined using this technique than with most other methods because heat fixation is not employed, and therefore little shrinking is encountered. However, since cells are normally colorless in the living state, little or no internal structure is seen using this staining procedure.

EXPERIMENT 2-2

Negative Staining of Microorganisms

MATERIALS

1. Mixed broth suspension of microorganisms, rods and cocci.
2. Dorner's nigrosin stain or fresh India ink.
3. Loeffler's methylene blue stain.
4. Xylene.
5. Congo red, 25% solution.
6. HCl, 1% solution.

PROCEDURE: Following aseptic technique, remove 2 or 3 loopfuls of the microbial suspension provided and deposit them on a clean slide. Add one loopful of the nigrosin stain to the microorganisms on the slide and mix the two suspensions with the loop, using a rotary motion. Spread the film out quite thin, particularly at the edges. It may be necessary to spread the film over most of the slide. With proper dilution the slide should appear gray rather than black. If the mixture is too thick, large cracks will appear in the nigrosin under the microscope. Allow this mixture to dry in the air; do not fix with heat. Examine the slide under the oil-immersion objective, and make a drawing of a representative field.* In addition, make a simple smear of this same suspension, fix with heat, and stain with Loeffler's methylene blue. Try to note any differences in the sizes of the same organisms, comparing the stained and unstained slides. On what is the difference based?

NOTE: Since it is often helpful for you to be able to refer back to previous work, a few permanent preparations may be made of your best efforts. Slide preparations can be made permanent by a very simple procedure. After examining the prepared slide to ascertain that it is good, remove the immersion oil by running xylene across it. Air dry. Place a drop of Canada balsam or Permount on the smear. Carefully place a cover slip (No. 1 thinness) over the smear and gently press until the Canada balsam has completely "filled in" under the cover glass. Maintain gentle pressure with your fingers for 2 to 3 min. Put the slide in your drawer to dry for about two weeks. After drying, scrape off any excess dried balsam from the edges of the cover glass. Many students want to make a permanent collection of this type for future reference.

---

*Another method which is useful employs the acid dye Congo Red. Place a drop of Congo Red at one end of another clean slide, add a loopful of microbial broth suspension to it and mix with a rotary movement of your loop. Now place one end of another slide in the dye-microbe drop at a 45° angle until the drop spreads to the edges of the glass, then slowly push the inclined slide at the same angle to the other end of the bottom slide. The dye should follow your movement, depositing a thin film across the slide. Air dry – do not fix with heat. When dry, flood the slide with 1% HCl and pour off immediately and do not wash. After the slide has air-dried, examine with the oil-immersion objective and make a drawing of a representative field.

A technique frequently used to determine the ability of an organism to move independently is the *hanging drop* procedure. In this preparation, organisms are seen in their natural living state, and if they are capable of independent movement, this can be observed easily. Microorganisms are customarily so small that when suspended in a fluid medium they are subject to movement that does not originate from the organism itself. As you know from your study of chemistry, all molecules are in constant motion. Microorganisms are small enough that collisions with the moving water molecules and other molecular-sized particles in solution cause the microbial cell to rebound slightly from the impact. Since there are many billions of molecules colliding with the microorganisms simultaneously, one observes that the cell bounces around in a rather indeterminate fashion. The cell does not appear to be going anywhere other than to "dance around" in one general vicinity. These organisms are said to exhibit *Brownian movement,* but it is important to realize that organisms exhibiting only Brownian movement are *not motile*. On the other hand, many microorganisms have organs of locomotion, which make it possible for them to move "purposefully" when in a fluid medium. Such organisms are observed to twist and turn and often go rapidly for a considerable distance across the microscope field. This type of movement is *true motility*.

EXPERIMENT 2-3

Motility Studies on Microorganisms

MATERIALS

1. Broth cultures of Micrococcus luteus* and Bacillus subtilis.
2. Hay infusion.**
3. Concave slide and vaseline.
4. Methylene blue (1:10,000) or aqueous crystal violet (1:2000).

PROCEDURE: This procedure is better demonstrated than described, so the instructor will give you a preliminary demonstration of this technique. Briefly, the steps are as follows: (a) a thin layer of vaseline is applied around the edge of the well in the concave slide, (b) a loopful of one of the microbial suspensions is put in the center of the cover slip, (c) the concave slide is inverted and pressed down firmly on top of the cover slip so that the cover slip completely covers the well, (d) the entire preparation is quickly turned right side up again, and (e) the edges of the cover slip are checked to make certain the vaseline has sealed properly.

---

*Micrococcus luteus is a round bacterium that is found in dairy products as well as in the air. It is a common contaminant of laboratory experiments and produces a yellow-colored colony.

**A suitable hay infusion can be prepared by putting grass, leaves, dirt, etc. into a container of water and letting it stand at room temperature for 3-4 days. During this time it is necessary to bubble air through the infusion, otherwise a layer of *Bacillus sp.* will form over the top and effectively suffocate most other organisms in the infusion.

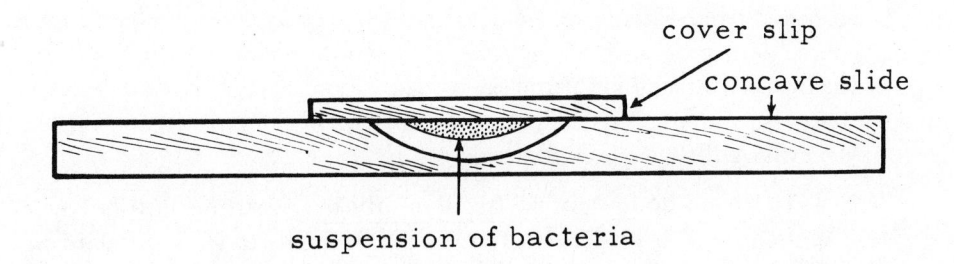

```
                    cover slip
                    concave slide
  ┌─────────────────────────────────────┐
  │                                     │
  └──────────────╲_____╱──────────────┘
                    ░░░
                    │
            suspension of bacteria
```

If you follow this procedure, your drop of microbial suspension will now be hanging on the under side of the cover slip within the air space created by the concavity of the slide. Place this slide preparation on your microscope stage so that the edge of your microbial suspension is located approximately in the center of the field under low power. Reduce the amount of light with the iris diaphragm (or by lowering the condenser) until the field starts to become dull. Then swing the high-dry objective into place without moving anything else. The suspension should nearly be in focus. Observe for microorganisms in motion. Remember they are quite small and they will appear very ghostlike under these conditions, because they have no natural color of their own and are translucent. During your observations, note which organism is motile and which organism is nonmotile. Repeat this procedure using a drop of hay infusion, which contains a wide variety of protists.

Now try a variation of the technique described above, using a loopful of methylene blue (diluted 1:10,000) or crystal violet (1:2000) mixed with a loopful of the bacterial suspension. This dilution of the dye gives a faint color to the organisms without affecting their viability. If you have had difficulty observing the organisms in the first procedure, this should make it easier for you. In the future, this latter technique is best to follow when you wish to use either a hanging drop mount or a wet mount for motility.

We now turn our attention to the study of the internal structure of microorganisms by means of both simple and differential staining. Many microorganisms contain small granules and vacuoles composed of a variety of materials. Such granules are probably stored metabolites, and are analogous to the stored fats and starches in higher forms. In order to visualize these structures, one must apply reagents that both react specifically with the substance one desires to see and give it a color that can be visualized. For example, it is well known that some microorganisms manufacture and store fat. Therefore, if one simply selects a dye that stains only fatty materials, a procedure for visualizing stored fat within microbial cells is obtained. In the subsequent experiment you will use such a dye (Sudan black B). In order to make fat globules stand out in greater relief, a red water-soluble dye (safranin) is applied to impart a different color to the remainder of the cell.

CYTOLOGY OF MICROORGANISMS

EXPERIMENT 2-4

The Demonstration of Fat Vacuoles

MATERIALS

1. A 24- to 48-hour slant of Bacillus subtilis grown on nutrient agar containing 5% glycerol (glycerol agar); a 24- to 48-hour slant of Saccharomyces cerevisiae* (baker's yeast) grown on malt extract agar.
2. Sudan black B stain.
3. Aqueous safranin (0.5%).
4. Xylene.
5. Lugol's iodine solution or Gram's iodine.
6. Blotting paper.

PROCEDURE: Prepare a smear of Bacillus subtilis on a clean glass slide. Allow to dry in air and fix with heat. Flood the slide with the Sudan black B and allow to react for at least 10 min. Do not let the stain evaporate while reacting. Drain off the excess dye and dry by placing a piece of blotting paper directly on the smear until all the dye soaks into the paper. Do not move the paper during this process. Carefully lift the paper off when the blotting is complete. Since this dye is selective for fat-soluble particles, only that type of substance is stained thus far. Add several drops of xylene to the slide to clear the excess stain, then blot dry again. In order to stain the remaining material in the cell, apply the aqueous safranin solution as a counterstain, and permit it to react for about 10 to 15 sec. Wash immediately with tap water and permit to dry in the air. Using the oil-immersion objective, look very carefully for the presence of blue-black globules of fat within the red cytoplasm. Small droplets of Sudan black may adhere to the glass, but this is not fat-staining. Make a drawing of what you see.

    Another way of observing the same substance can be shown using another organism. On a clean slide, mix a loopful of Sudan black B, a loopful of Saccharomyces cerevisiae culture, and a loopful of tap water. Carefully place a cover slip over the preparation and examine with the high-dry objective and also with the oil-immersion objective. It sometimes takes a few minutes for the dye to react with the fat globules, so continue your observations for at least 10 min. Make a drawing of what you see. This kind of a preparation is known as a wet mount and is commonly used in cytological studies.

    Sometimes it is also possible to observe deposits of glycogen in yeast cells. For this purpose, make a wet mount of S. cerevisiae,

---

Saccharomyces cerevisiae is not a bacterium, but a member of the mold group of organisms. It is generally egg-shaped, somewhat larger than the usual bacterium, and is well-known for its production of ethyl alcohol and carbon dioxide as end products of its metabolism.

using Gram's iodine or Lugol's solution, and observe for red-brown deposits darker than the normal brown tinge imparted by the iodine. Make a drawing if you find some deposits.

In the subsequent experiment, you will attempt to visualize granules that have been given a variety of names by different investigators in the past. They are called *metachromatic granules,* due to the fact that some dyes are observed to change color in its presence. They have also been called *volutin* granules and probably have other names as well. These granules are composed of polymerized inorganic phosphates which represent a stored form of this element, widely used in various metabolic processes in the cell. These granules become markedly depleted during starvation conditions. Metachromatic granules have a very strong affinity for basic dyes; if a basic dye is applied to an organism containing such granules, they will be stained much more intensely than the remainder of the cell.

EXPERIMENT 2-5

The Demonstration of Metachromatic Granules (Volutin)

MATERIALS

1. Young agar slant culture (18 to 24 hours) of Corynebacterium pseudodiphtheriticum.*
2. Albert's diphtheria stain.
3. Lugol's iodine solution.
4. Loeffler's methylene blue.
5. Hay infusion.

PROCEDURE: Make sure that your slides are well cleaned for this exercise. Prepare a smear of Corynebacterium pseudodiphtheriticum in the usual fashion. Use very gentle heat fixation of the organisms in this experiment. Flood the smear with Albert's staining solution and allow the stain to react for about 5 min. It sometimes helps to apply a very little heat to the stain itself. This is done by inverting the lit Bunsen burner over the slide so the flame touches the stain deposited upon the slide. Do this only briefly. Do not allow the stain to steam or boil. After 5 min. drain the slide to remove excess dye but do not wash in tap water. Apply the Lugol's iodine solution to the wet slide and allow this to remain for about 1 min. Wash the slide in tap water to remove the iodine and allow it to dry in the air. Examine this preparation under the oil-immersion objective and look for metachromatic granules within the cytoplasm of the organisms. They

---

*Corynebacterium pseudodiphtheriticum is a small slender rod found as a normal nonpathogenic inhabitant of the nose and throat of humans. It is also known as Hofmann's Bacillus.

should appear as very intense blue round dots within a pale green stained cytoplasm. Often with this stain the cytoplasm is very faint and hard to distinguish, whereas the granules usually show up quite clearly. Again, only granules within the cytoplasm are significant. Droplets on the glass can be confusing. It may require some critical focusing and light adjustment to see the entire cell properly.

Another method for showing the metachromatic granules is to make a smear as in the first part of this experiment and stain it with Loeffler's methylene blue for 5 min. Wash the stain off with tap water, allow to dry in air, and examine the slide under the oil-immersion objective. The granules again stain a very intense blue color, whereas the cytoplasm takes on a much paler blue color. Make appropriate drawings on your report sheet for each staining method.

Using the Albert's method, stain a smear from the hay infusion provided. Many times these infusions contain long spiral rods (spirilla) that are very rich in metachromatic granules. Make drawings under oil-immersion.

It should now be obvious to the student that although bacteria are exceedingly small, one is able, with appropriate techniques, to show internal structures. The stains that have been used up to this point are primarily concerned with the identification of certain specific substances within the cell. Now we shall turn our attention to another staining technique using two of the best known differential stains.

The most widely used bacteriological stain is the differential *gram stain,* named after Dr. Christian Gram, the originator. Another stain widely known to microbiologists is the *acid-fast stain.* Both stains are used to distinguish one group of organisms from another. All cells can be classified into at least two categories using either stain, since both stains may show a positive or negative result. The gram stain is an extremely useful classification tool. One immediately divides the entire microbial kingdom into two categories when the gram stain is applied. Furthermore, by observing the shape of the cell involved, two or more additional divisions can be made. Therefore the gram stain actually permits a division of the microbial kingdom into a least four distinct parts.

The acid-fast stain may be used similarly, but here the actual result is not quite the same. Although one could categorize all organisms as acid-fast or not (either rods or cocci), in practicality this is not true because most organisms are *not* acid-fast — that is, they do not react positively to this stain. Only a small group of microorganisms are acid-fast, and this stain therefore is quite useful for identifying this group (mostly members of the genus *Mycobacterium).*

In some instances, the gram stain can be used for tentative identification, but in most cases it cannot. The gram stain is usually the beginning point in an identification procedure, whereas the acid-fast stain is frequently used for positive identification, since so few organisms are acid-fast.

The gram stain is composed of four different reagents, and a certain amount of skill is necessary to make the stain work correctly. Crystal violet (also known as gentian violet) is ordinarily the first or *primary dye.* Its purpose is to impart color to *all organisms* in the smear. The second reagent is a diluted iodine solution, which is normally referred to as Gram's iodine or occasionally Lugol's iodine. This solution acts as a *mordant* — a substance that enhances or strengthens the union between a dye and its substrate. The third reagent is ordinarily referred to as a *decolorizer* because it dissolves and removes the primary stain from the cells. Some laboratories use a mixture of acetone and 95% alcohol in proportions of 30% and 70% respectively, while others use only 95% alcohol. Crystal violet is soluble in this combination and will dissolve in it. Many organisms retain the primary dye with a great deal of tenacity despite decolorization. These organisms are called *gram-positive.* The fourth reagent in the gram stain is a red dye called safranin, which is applied as a *counterstain.* If an organism has been stained and then lost the stain through the decolorizing procedure, it will now be invisible (unstained), just as it was at the beginning. To make the organism visible, counterstain of a color different from the primary stain is applied. Any such organism that takes up the red safranin stain would be designated as *gram-negative.* However, if the organism retains the primary dye throughout the staining procedure, it is designated gram-positive. Such organisms do *not* take up the red safranin dye simply because they already have completely reacted with the crystal violet.

The essential difference between gram-positive and gram-negative cells appears to reside in the behavior of the cell walls during the decolorization step. Cell walls of gram-positive organisms are more sensitive to dehydration by alcohol, the latter substance acting to close up the pores of the wall. Under these conditions the crystal violet-iodine complex cannot escape the cell and thus the primary color is retained. Gram-negative cell walls, which have a much higher lipid content than gram-positive cell walls, apparently do not close up during alcohol treatment and the primary dye is free to leach out with the decolorizer. Therefore, staining a mixture of cells of different types with the gram stain results in some organisms that are blue and some that are red. Those species that are blue are gram-positive and those that are red are gram-negative.

It should be mentioned here that the gram-staining character of many species is variable. Further, all gram-positive bacteria do not retain the primary dye to the same extent and thus a gradient of gram-positivity exists. For these reasons and others, the proper performance of this stain requires much skill in order to be confident of the results.

It would be well to mention here some observations regarding pitfalls in the performance of the gram stain. Decolorization is the most critical step in the gram stain. It is not an all-or-none type of procedure. If you decolorize too vigorously, *all* cells will lose their primary dye and thus all organisms will appear to be gram-negative. On the other hand, if you decolorize too gently, even gram-negative organisms will retain the primary dye and be seen as gram-positive. There is no easy way to describe a correct decolorization procedure. The best guide is one's own experience. When you have done several of these gram stain procedures and have checked yourself for accuracy, you should get a reasonable idea of how much decolorization is necessary.

The final result of the gram stain is also affected by the *age of the culture*. It is exceedingly important for the student to remember that the gram stain as a differential procedure is valid only with cultures that are *24 hours old or younger*. Once the culture is older than 24 hours many gram-positive cells begin to lose the ability to retain the primary dye. Aging of the cells tends to make them become gram-negative even though they really are not. Therefore, the gram stain should be done only with young cultures of microorganisms.

EXPERIMENT 2-6

The Gram Stain

MATERIALS

1. 24-hour cultures of Escherichia coli* and Staphylococcus epidermidis**, and a 48-hour culture of Saccharomyces cerevisiae (baker's yeast).
2. Gram's crystal violet stain.
3. Gram's iodine solution.
4. Acetone-alcohol or alcohol (95%) decolorizing solution.
5. Gram's safranin.
6. Applicator stick.

PROCEDURE: Clean four glass slides and prepare four identical smears on each slide in the following fashion: near one edge of the slide (leaving a small area for holding the slide), prepare a small smear of E. coli about the size of a dime. Just to the right of this first smear, prepare another smear of the same size with Staphylococcus epidermidis. Then prepare a third smear next to the second one with the yeast suspension, Saccharomyces cerevisiae. Finally, nearest to the outside edge of the slide, prepare a small smear using material scraped from your gum line with an applicator stick and previously emulsified in a drop of water. With a wax pencil label the slides 1, 2, 3, and 4. Allow all four slides to dry in the air, and then fix with heat in the usual fashion. Flood slide No. 1 with Gram's crystal violet (gentian violet) making sure all smears are covered and allow it to react for 1 min. Pour off the excess stain and wash briefly in tap water. Apply Gram's iodine and allow it to react for 1 min. Wash off with tap water. Now carefully add the decolorizing solution, one

---

Escherichia coli is found as a normal inhabitant of the intestinal tract of man and a variety of other vertebrates. It is a small, short rod and grows very rapidly in laboratory cultures. Although some stains are pathogenic, most of them are not.

**Staphylococcus epidermidis* is a round bacterium that grows typically in clusters that resemble a "bunch of grapes." It is a normal inhabitant of the human skin and is occasionally found as the cause of minor skin abscesses.

drop at a time to each of the smears, with the slide tilted over the sink. As soon as color stops coming off the slide, rinse it in tap water to stop the decolorizing action. Finally, flood the slide with Gram's safranin solution and allow it to react for 10 to 20 sec. Wash off the excess dye and allow it to dry in the air. Repeat the same procedure with slide 2, but do not counterstain with safranin. In other words, stop the procedure after the decolorization. Allow this to dry in the air. Repeat the procedure with slide 3, but this time stop before decolorization. In other words, do not decolorize this slide. Allow to air dry. Do the first step only with slide 4. Wash and allow to dry. Using the oil-immersion objective, observe the slides, and make a drawing showing a few organisms from each of the 16 smears you have prepared, indicating what color, if any, the organisms display.

The procedure you employed on slide No. 1 is the complete gram-staining procedure. When you use this stain you will ignore the procedure for slides 2, 3, and 4. These were only added to this particular exercise to show you the appearance of the cells at each step of the staining procedure. When you are asked to do a gram stain in future experiments, it is procedure 1 that you will follow.

The acid-fast stain is so named because of the chemical nature of the decolorizing agent — a solution of 3% HCl in 95% alcohol. This is a much more intensive decolorizer than the acetone-alcohol used in the gram stain, and care should be taken not to use acid-alcohol in the gram stain. Acid-fast organisms are capable of retaining the primary stain through a very intensive decolorizing procedure. The retention of the primary dye by organisms which can do so in this state appears to be based on the fact that the primary dye is more soluble in the lipoidal substances found in these organisms than the dye is in the decolorizing solution. In the acid-fast stain, therefore, decolorization is not nearly as delicate a procedure as in the gram stain. However, acid-fast organisms that have been grown in artificial culture for some time tend to become less strongly acid-fast than those freshly isolated from their natural habitat.

Another feature of the acid-fast stain is the use of heat as an *intensifying agent* for the primary stain. The primary stain is literally driven into the bacterial cell by heating it to steaming. This procedure is necessary because members of the genus *Mycobacterium* (which are generally acid-fast) possess a very high complement of fatty materials in their cytoplasm. Therefore, either heating or some form of surface-active agent (such as a detergent) is necessary to make the primary dye penetrate well. An alternative procedure is to mix the primary stain with a strong detergent before applying it to the smear. Such a solution does not need to be heated in order to penetrate. Carbolic acid (phenol) is added to the carbolfuchsin (primary dye) as a chemical intensifier to further assist in penetrating the cell.

The acid-fast stain finds a great deal of application in medicine, for some microorganisms in the genus *Mycobacterium* are associated with infection, and there are very few other bacteria that give a positive acid-fast stain.

Mycobacterium tuberculosis is the causative agent of tuberculosis. There are other genera of bacteria that are either partially or totally acid-fast, but they are less commonly associated with human disease. In general, acid-fastness is very rare in any cell.

EXPERIMENT 2-7

The Acid-Fast Stain

MATERIALS

1. A 72-hour culture of Mycobacterium tuberculosis* (avirulent strain). When available, a specimen of autoclaved tuberculous sputum will also be provided. A 24-hour culture of Staphylococcus epidermidis.
2. Ziehl-Neelsen's carbolfuchsin stain.
3. Loeffler's methylene blue stain
4. Acid-alcohol.
5. Egg albumin solution (1.0%) in saline.

PROCEDURE: Prepare a mixed smear of Mycobacterium tuberculosis and Staphylococcus epidermidis by emulsifying the growth in some 1% albumin solution. The albumin solution helps to keep the Mycobacteria tuberculosis adherent to the slide, since they would normally tend to wash off easily. Their large fat content undoubtedly accounts for this behavior on slides. If autoclaved tuberculous sputum is available, make a second smear of that material alone. Fix those preparations with heat after drying in the air. Suspend the slides over the sink with slide holders and flood them with Ziehl-Neelsen's carbolfuchsin stain. Grasp the bottom of the Bunsen burner and pass the flame back and forth across the top of the stained slide until the steam just begins to rise. Do not boil this stain; merely keep it at the steaming point and continue this for 5 min. If the stain evaporates at the edges, replenish with additional stain. Wash the stain off with tap water and decolorize the preparation by adding acid-alcohol one drop at a time to the smear until all color stops coming off the slide. Wash the slide in tap water and counterstain with Loeffler's methylene blue stain for 1 min. Wash the slide with water and air dry. Examine the preparation under the oil-immersion objective and make a drawing of a typical field for each preparation.

---

*Mycobacterium tuberculosis exists in both pathogenic and nonpathogenic strains, the latter differing in certain respects from the former. They are fairly long, very slender rods.

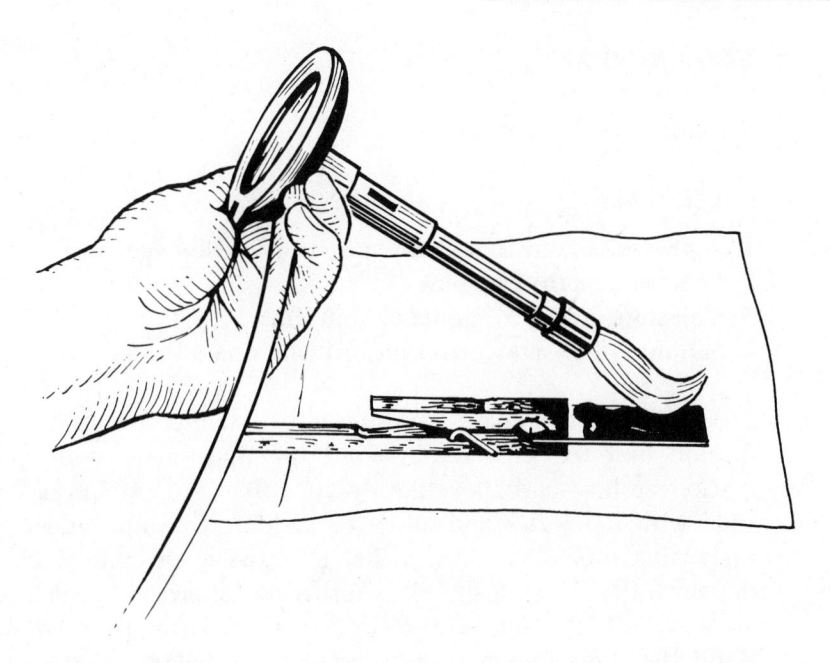

Figure 2-2 Flame the stain to steaming by holding the burner as shown above, moving back and forth the width of the stain. Occasionally remove the flame to observe for rising steam, the flame intermittently to maintain the stain at this temperature for the prescribed time interval.

Now that you have had some experience with the use and application of differential staining for classification purposes, the next step is the study of specific structures of the cell by means of differential staining. As you know, even though microorganisms are exceedingly small, structures can be observed within and attached to such cells provided proper staining techniques are employed. Outstanding in this respect are endospores, flagella and capsules. Also, the bacterial nucleus and cell wall have been seen with the help of special stains.

Making one or more of these structures visible involves taking advantage of some peculiarities of the structure involved. For example, the bacterial endospore, once formed, is exceedingly resistant to staining by the usual methods, and this characteristic suggests one way to make this structure visible. The application of a simple stain to a bacterial cell that contains an endospore imparts color to the nonspore portion of the cell *(vegetative cell)* while the spore itself remains uncolored, therefore distinguishing it. Other staining methods use intensive measures to drive the stain into the endospore. Once the stain is in the endospore, it can be removed only with equally intensive decolorization measures, and decolorizing procedures can be used that would destain other parts of the cell quite easily, leaving the spore still stained. To make the distinction clear between the spore and the vegetative portion of the cell, a contrasting counterstain is usually applied in the ordinary fashion and the resulting picture shows the initial stain taken up by the spore and the second stain appears in the cytoplasm. Thus it makes for a very simple method of distinguishing the endospore from the vegetative cell.

EXPERIMENT 2-8

Demonstration of Bacterial Endospores

MATERIALS

1. A 48- to 72-hour agar slant culture of Bacillus species.
2. Loeffler's methylene blue.
3. Malachite green (5% aqueous solution).
4. Safranin (0.5% aqueous solution) (not Gram's safranin).

PROCEDURE: Prepare two smears of Bacillus species in the usual fashion and fix with heat. Flood the first smear with Loeffler's methylene blue and allow the dye to react for 3 to 5 min. Wash the slide with tap water and allow to air dry. Examine under the oil-immersion objective and make a drawing of the field, which demonstrates the endospores. Another useful method to demonstrate endospores is the Schaeffer-Fulton method. Using the second smear, flood the slide with malachite green stain. Allow the stain to react for about 1 min., then heat the slide until it steams for another minute. Wash the slide in tap water for about ½ min. Counterstain the smear with 0.5% safranin solution (do not use Gram's safranin), and allow this stain to react for ½ min. Wash the slide in tap water and allow to air dry. Examine under the oil-immersion objective, and draw spores and vegetative cells. Sometimes it is necessary to look at several parts of the slide before you can find a field of view that sufficiently demonstrates the structures. The spores are green and vegetative cells are red.

A somewhat difficult problem confronts us when we attempt to visualize the bacterial capsule. The capsule is ordinarily composed of water-soluble material. Therefore, if we use the previously discussed procedures, the capsule is very frequently washed away. Furthermore, the capsule is usually composed of nonionic substances, usually polysaccharides. As you know, the staining technique itself is frequently based upon chemical interaction between ionized substances. Actually the best method for visualizing capsules combines use of the phase microscope and material in the medium surrounding the capsule that will give a background to set it off. You can duplicate this on a modest scale without phase equipment by using a wet mount procedure in which a few carbon particles are put in the suspension along with the encapsulated bacteria. By reducing the light intensity with the iris diaphragm (or lowering the substage condenser) you can get a phaselike effect and can see the capsules as halos or colorless rings around the organisms. Sometimes stains are successfully employed in which the capsular material is first dehydrated with a solution of highly concentrated salt; this procedure does not work well in the student laboratory and is not used here.

EXPERIMENT 2-9

Demonstration of Bacterial Capsules

MATERIALS

1. 24-hour cultures of Klebsiella pneumoniae (or Aerobacter aerogenes)* and Staphylococcus epidermidis on tryptose-phosphate agar slants.
2. India ink (freshly filtered).
3. Loeffler's methylene blue.

PROCEDURE: Prepare a very dilute suspension of Klebsiella pneumoniae (or Aerobacter aerogenes) on the slide by using several drops of tap water. Place the least amount of growth on the slide that you can retrieve from the culture by touching the growth with the edge of the inoculating loop and emulsifying this gently with the water on the slide. To this diluted suspension, add India ink, barely touching the edge of the loop to the India ink and then mixing the ink with the bacteria and water suspension. If you do this properly, the suspension will be dark gray after thorough mixing. Carefully place a cover slip over this preparation and examine with the high-dry objective. The black particles of India ink are seen by their rapid Brownian movement, and the bacteria are pale, ghostlike forms. When you bring a bacterial cell into critical focus, reduce the light with the iris diaphragm until the cell becomes somewhat darkened. If the preparation has been made correctly, you should be able to see a halolike effect around the organism that represents the capsule. Note that the rapidly moving carbon particles bounce off of the edge of this halo and do not actually strike the bacterial cell. Thus you can actually visualize the capsule of this organism. After observation under high-dry, very carefully place a drop of oil on the cover slip and examine under oil immersion. Draw a typical field when you have achieved critical focus using this magnification.

Repeat this procedure with the Staphylococcus epidermidis culture. Careful examination should reveal the particles of carbon actually striking the bacterial cells, because there is no intervening capsule to ward the particles off. You should repeat this experiment several times to acquire skill in the procedure, because you will use this technique on an "unknown" later on.

A simple modification of this technique is Gin's method. Mix equal amounts of India ink and water, then add a loop of K. pneumoniae culture and gently spread it. Allow the smear to dry

---

Klebsiella pneumoniae is very closely related to *Aerobacter aerogenes;* so much so, in fact, that it is very difficult to distinguish between them. It is a rare cause of pneumonia in man and apparently can be isolated as a normal inhabitant of human mucous membranes. It is a very short, stubby, rod-shaped organism (coccobacillus) which is known to produce a very thick gelatinous capsule.

<u>without</u> heat fixation. Counterstain with methylene blue for 3 min.; rinse and dry. Examine under oil-immersion and draw.

Another procedure is used to stain flagella. This requires a great deal of laboratory skill and is one which is somewhat difficult to employ in the student laboratory. The theory of this procedure is rather simple. The stain is applied along with a tannic acid mordant that sticks not only to the cell but to the flagella in successive coats until the bacteria and flagella are actually made larger than normal. This is necessary because flagella are normally too thin to be seen with the optical microscope. This material, combined with the stain, gives a picture of the cell and flagella in a somewhat enlarged and distorted state. You will utilize this procedure using a commercially prepared stain.

There are many pitfalls involved in using the flagella stain, the most notable of which is the presence of extraneous organic matter upon the slide itself. The mordant has a very powerful affinity for virtually any kind of organic matter, and it will stain all kinds of debris that might be present. To avoid this, the slide used for the flagella stain must be thoroughly cleaned beforehand. Preferably the slides should be new, and once put into use, they should not be handled except at the edges. Another problem in the use of flagella stain is that the flagella themselves are exceedingly fragile and tend to break off very easily. Thus, with any flagella stain preparation, you customarily see many free flagella scattered on the slide, unattached to cells. However, patience and careful attention to detail usually result in an excellent preparation.

The use of the bacterial nuclear stain procedure involves the use of a highly dangerous chemical and is not attempted in the student laboratory. If possible, a demonstration slide will be provided so you may see the bacterial nucleus through the demonstration microscope.

EXPERIMENT 2-10

The Flagella Stain

MATERIALS

1. A distilled water suspension* of a very young actively motile culture of <u>Proteus vulgaris</u>** (6 to 12 hours).
2. Chemically clean slides (soaked in chromic acid for 24 hours and rinsed in distilled water). These will be distributed to you individually.
3. Leifson's flagella stain (freshly filtered).
4. 1 ml. sterile pipet (1).

---

*Gently wash the surface growth from nutrient agar slants with distilled water and transfer suspension to clean tubes.

***Proteus vulgaris* is a small rod-shaped bacterium that is normally found in putrifying materials in nature, and may also be found as a member of the normal flora of the human intestinal canal. Most strains are very actively motile.

PROCEDURE: Prepare a slide for staining by outlining about two-thirds of the slide with a heavy wax pencil mark so as to form a completely enclosed rectangle:

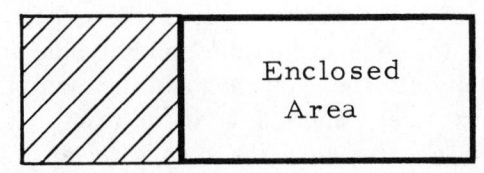

Enclosed Area

Use the nonenclosed area for handling the slide. Remove with a pipet 0.1 ml. of the bacterial suspension from the surface portion of the water suspension of Proteus vulgaris (most of the motile organisms are found here) and put one drop of this suspension near one end of the enclosed area. Tilt the slide at a 30 to 40 degree angle and allow the drop of suspension to run down the slide to the other end. If the slide has been properly cleaned, the drop will flow in a straight line down the glass. Leave the slide in a tilted position and allow to air dry. Do not fix with heat. Flood the slide with Leifson's stain, completely filling the enclosed area. Allow the stain to react for 10 min. Shake off the excess dye and wash gently in tap water. Allow to dry in the air and examine under oil-immersion. You will have to search to find a field of view in which clear examples of flagellated organisms can be found. It is a peculiarity of the flagella stain that much of the preparation is not clear. Ordinarily there is a portion of the slide in which flagella clearly appear.

The study of other fine structures of microorganisms is often greatly aided by staining techniques. For example, the bacterial cell wall can be visualized very easily in this way, even though it does not appear as a distinct entity on the usual stained preparation. In using cell wall stain, advantage is taken of the fact that the cytoplasmic contents can be differentially hydrolyzed by phosphomolybdic acid and the cell wall left intact. Appropriate staining reveals the cell wall with a "hollowed out" cytoplasm. The cell wall can also be visualized indirectly by taking advantage of the fact that certain simple stains (e.g., methylene blue) do not stain the cell wall while others (e.g., crystal violet) stain the cell wall as well as the cytoplasm. This is why the same cells stained with methylene blue appear smaller than those stained with crystal violet.

EXPERIMENT 2-11

Demonstration of the Bacterial Cell Wall

MATERIALS

1. Young slant culture of Bacillus subtilis (12 to 18 hours).
2. Phosphomolybdic acid solution (1.0%).
3. Aqueous solution of methyl green (1.0%).

PROCEDURE: Make a relatively thick smear of Bacillus subtilis on a clean glass slide. While the material on the slide is still wet and unfixed, add the phosphomolybdic acid solution directly to the smear until it is entirely flooded. Allow the acid solution to react 4 min. Tilt the slide so the phosphomolybdic acid solution drains off completely into the sink; then add the methyl green stain directly to the moist smear and allow it to react 5 min. When staining is completed, gently wash the slide with tap water. Be careful not to remove any more of the growth than is absolutely necessary, but do a complete job of washing, making certain to remove all excess stain. Allow the stain to dry in the air; then examine under oil-immersion. Properly stained cell walls take on a dark green or blue color, and the cytoplasm is either unstained or very pale green. You will have to search the slide to find the best representatives when you use this particular stain. Draw your observations.

Although the technique for visualizing the bacterial nucleus is somewhat impractical for use in the student laboratory, it is quite possible to show the nucleus of ordinary yeast cells by staining. Some of the same difficulties are encountered with yeast as with bacteria, but are more easily resolved. Certain cytoplasmic constituents (that normally interfere with staining the nucleus) must be hydrolyzed before adding the nuclear stain (toluidine blue). The general morphology of typical yeasts can also be observed in the next experiment.

EXPERIMENT 2-12

Demonstration of the Nucleus and General Morphology of Yeasts

MATERIALS

1. Young culture (6 to 12 hours) of Saccharomyces cerevisiae on Sabouraud agar slants.
2. Ethanol solution (40%).
3. Potassium hydroxide solution (0.1 M).
4. 0.1% toluidine blue in 10% ethanol.
5. Ethanol solution (10%).
6. Gram's iodine solution.

PROCEDURE: Prepare a semidense smear of the yeast cells in tap water and allow to air dry. Very gently fix with heat. Flood the slide with 40% ethanol solution for 1 min. and wash with tap water. Flood the smear with KOH solution and allow to react for 1 hour. If dehydration occurs, add more KOH. Wash well with tap water, then apply the toluidine blue stain for 2 min. Wash the stain off with the 10% ethanol solution, then shake off all excess moisture and air dry.

CYTOLOGY OF MICROORGANISMS   49

Prepare a second slide in the same fashion down through treatment with 40% ethanol solution. Omit the KOH treatment. Apply the toluidine blue stain for 2 min. and proceed as with the other smear. Compare the two slides. The second slide shows that prehydrolysis of cytoplasm (KOH treatment) is necessary before applying the stain. Examine the preparations under oil-immersion and draw them. The nuclei stain a dark blue or pink while the cytoplasm appears paler and somewhat pink in color.

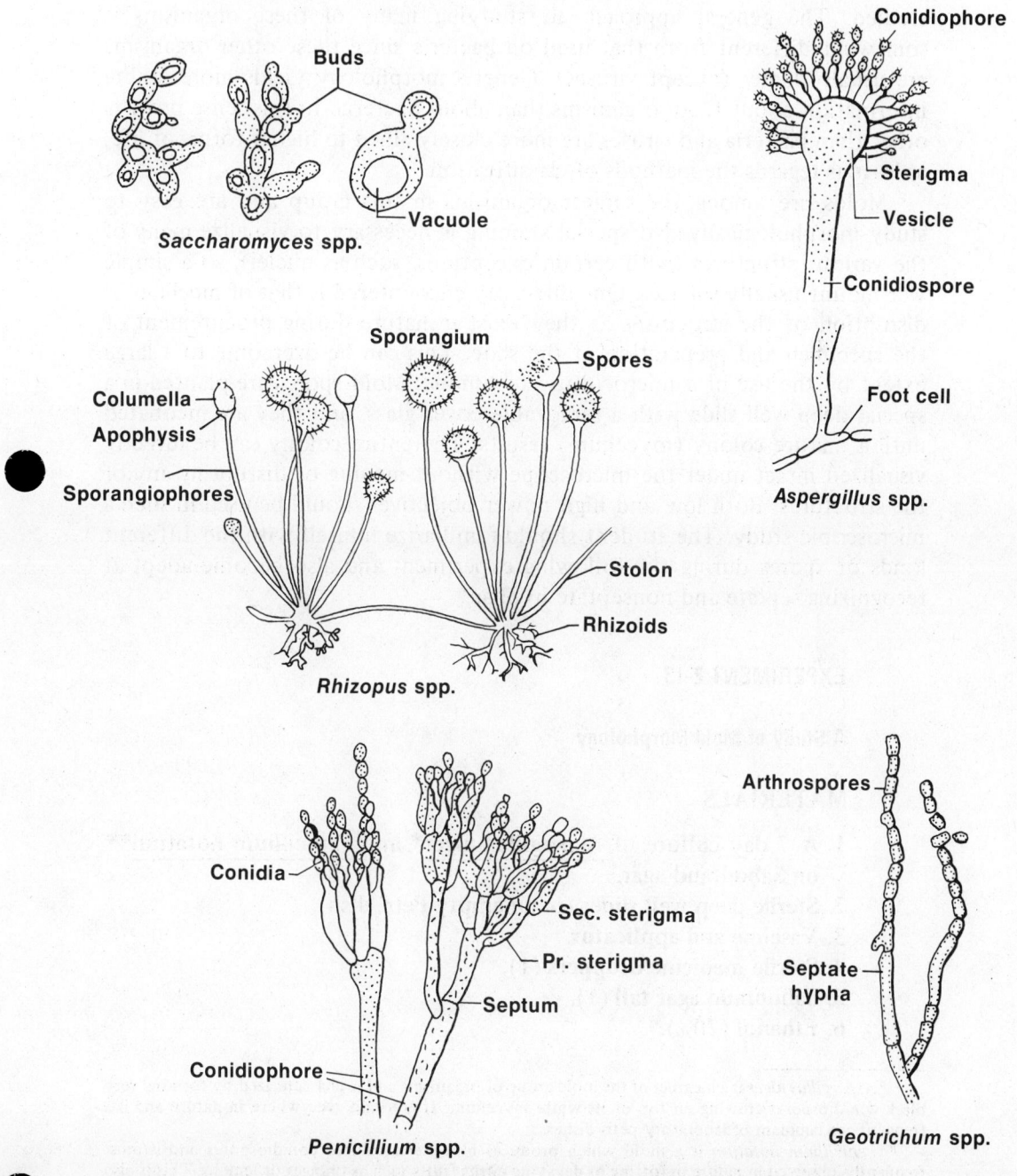

Figure 2-3 Vegetative structures of some typical fungi: yeast and molds.

In order to study additional yeast morphology, prepare a wet mount using Gram's iodine instead of water. Look for evidence of budding and internal structures, such as a large vacuole. Draw representative cells, using both high-dry and oil-immersion objectives.

Microbial cytology is also concerned with the study of protists other than bacteria. Molds, yeasts, and protozoa are frequently included for study in beginning microbiology laboratory, while algae and viruses are less often studied. The general approach to studying many of these organisms is somewhat different from that used on bacteria since these other organisms are usually larger (except viruses). General morphology yields more usable information about these organisms than about bacteria. In this sense protists other than bacteria and viruses are more closely allied to higher forms of life, at least as regards the methods of classification.

Molds are among the largest organisms in this group and are easy to study morphologically. No special staining is necessary to visualize many of the various structures (with certain exceptions, such as nuclei), so a simple wet mount usually suffices. One difficulty encountered is that of mechanical disruption of the structures as they exist in nature during procurement of the specimen and preparation of the slide. This can be overcome to a large extent by the use of a microculture technique. Mold spores are planted in a special deep well slide with a removable cover glass, and they are incubated until a mature colony (mycelium) results. The entire colony can be directly visualized intact under the microscope without moving or disrupting any of the structures. Both low and high power objectives should be used in such a microscopic study. The student should familiarize himself with the different kinds of spores during the following experiment and also become adept at recognizing septate and nonseptate hyphae.

EXPERIMENT 2-13

A Study of Mold Morphology

MATERIALS

1. A 7 day culture of <u>Aspergillus niger</u>* and <u>Penicillium notatum</u>** on Sabouraud agar.
2. Sterile deep well slides (in an empty Petri dish).
3. Vaseline and applicator.
4. Sterile medicine droppers (1).
5. Sabouraud agar tall (1).
6. Ethanol (70%).

---

*Aspergillus niger is a member of the mold group of organisms and is characterized by forming very black conidiospores growing on top of its white mycelium. It is found everywhere in nature and is a frequent contaminant of laboratory petri dishes.

**Penicillium notatum is a mold which produces blue-green colored conidiospores and is most frequently observed in nature in rotting or decaying citrus fruits such as oranges or lemons. This is also the mold that is responsible for the production of the antibiotic penicillin.

Preparation of a Mold Microculture

PROCEDURE: Melt the Sabouraud agar tall and cool to 45°C. While this is cooling, gently warm the Petri dish (with the flame of your Bunsen burner or by leaving it in the 37°C. incubator for about 30 min.) containing the sterile deep well slide. Remove a slide and close the Petri dish for later use. Quickly apply a thin layer of vaseline around the edge of the well with a toothpick or applicator stick. Remove some cooled, melted Sabouraud agar from the tall with the sterile medicine dropper, and put 5 drops of it into the bottom of the well. Put the cover slip, previously dipped in alcohol (ethanol) for a few minutes and air dried, in place over the well so that the opening is completely sealed off; then stand the slide on its side and leave until the agar hardens (see diagram):

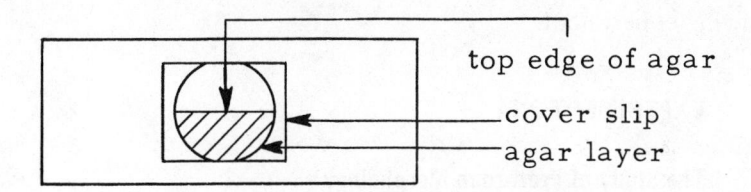

```
                              top edge of agar
                              cover slip
                              agar layer
```

After the agar has hardened, gently slip the cover glass down until the surface of the agar is accessible with an inoculating needle. Streak some of the spores from either of the mold cultures provided on the top edge (surface) of the agar and slide the cover glass back into the sealed position. Wipe off any excess vaseline around the cover slip, then incubate on its side in your drawer or in a 30°C incubator for five to seven days. When sufficient growth has taken place, examine the microcolony under low power by putting the slide on its side on the microscope stage. Make a drawing of typical hyphal strands as well as any spores. Label all parts appropriately. Exchange your culture with someone who has prepared the other organism and again make drawings as instructed above. Try to identify the following general structures: spores (type), sterigma, conidiophore, and hyphae (type).

Observation of Molds from the Air

Pour the unused Sabouraud agar tall into the unused sterile Petri dish from the first part of the experiment and allow to harden. Expose the plate to the air for 30 min. and incubate in your drawer for five to seven days. Tease a small portion of the mycelium from any two representative mold colonies with forceps and prepare wet mounts of each with tap water and cover slips. Make drawings of each using both low and high power to observe the various structures, labeling the type (and color, if present) of spores and hyphae.

CYTOLOGY OF MICROORGANISMS

The protozoa are traditionally classified as single-celled animals although they are also considered in the domain of the protists by many microbiologists. Most protozoa are quite large by microbiological standards and internal cellular structure is much more easily visualized than in the case of bacteria. Organisms in the class Sarcodina (e.g., *Amoeba proteus* and *Endamoeba histolytica*) ordinarily show nuclei, vacuoles, and other internal organelles in stained preparations. The class Ciliata (e.g., *Paramecium caudatum*) display even more internal structures, such as cilia, large and small nuclei, vacuoles, an oral groove, an anal pore, etc. The same is generally true of members of the class Flagellata and the class Sporozoa. Many of these organisms display various types of cell reproduction other than simple fission, and these reproductive methods can often be visualized, as is true of the process called *conjugation*. In many ways these forms show more relation to higher forms of life than do bacteria and yeasts. Experimentally it is more convenient to study these organisms from fixed and stained preparations, as in the following experiment.

EXPERIMENT 2-14

The Study of Protozoan Morphology

MATERIALS

1. Prepared slides of various protozoa, representing each of the four major classes: Amoeba proteus, Paramecium caudatum, Euglena viridis, and Plasmodium malariae.
2. Hay infusion.
3. Methyl cellulose (10%).
4. Prepared slides: Paramecium in conjugation and Paramecium in fission.

PROCEDURE: Examine each of the prepared slides assigned by your instructor and make careful drawings of selected cells from each. Use your own judgment as to which magnification should be used to visualize a given structure clearly. Make one drawing of each slide and label all parts of the organism clearly. Refer to your text or similar materials as a guide to the identification of the various organelles.

Examine a drop from the hay infusion by preparing an ordinary wet mount. Note the active motility of the protozoa. There is a technique for "slowing" them down so that you can study them more carefully. On a clean slide draw a circle of 10% methyl cellulose with a toothpick or applicator stick about the size of a dime. Place a drop of hay infusion in the center "well," and put a cover slip over the preparation. Examine with both low and high-dry objectives. The organisms that swim into the methyl cellulose become "stuck" and thereafter move only very slowly. Draw two different species observed.

```
┌─────────┬─────────┐
│/////////│    ◯    │  ← Methyl
│/////////│         │    cellulose
└─────────┴─────────┘
```

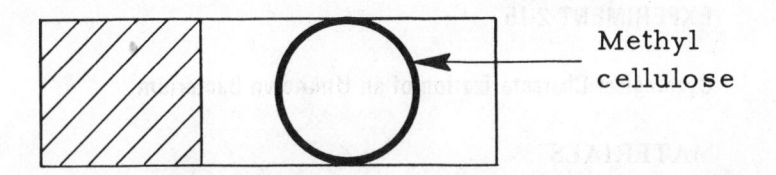

Cell division in protozoa can be studied best by using prepared slides in which the nuclear material is displaying <u>binary fission in Paramecium</u>. Examine this preparation and make a drawing that demonstrates this process. <u>Conjugation</u> is also more easily seen on a prepared slide. Again examine with the high-dry objective and draw.

Previous experiments have given you some idea of techniques involved in the study of microbial cytology. In order to firm up your knowledge in this area, it is usually desirable to go back over the ground just covered. These techniques are accepted tools in the classification of microorganisms, and any organism could be subjected to any one of these experiments as a means of aiding in its classification. If you were confronted with an unknown organism, one way to approach the problem of identification would be to make a thorough cytological study of that organism. This is precisely what the next experiment demonstrates. You will be given an unknown organism, and you will be asked to characterize the organism in terms of its cytology. This has certain advantages before you go to the next chapters. First of all, you will have a chance to repeat the procedures just completed to make sure that you have mastered the technique of making such a study. You will be told if you are correct or incorrect in each test performed. Secondly, repetition of these procedures should impress upon you that these stains have universal application in the microbial world. Finally, you will be on your own with an assigned task to test your skill in laboratory techniques.

Several things should be mentioned in connection with a cytological study of unknown microorganisms. One of the most difficult things for the beginning student to recognize in such a study is that a single species of bacteria does not necessarily look exactly the same from time to time or from one stain to another. For example, undoubtedly the most difficult problem of this type is the determination of whether or not a bacterium is a rod or a coccus when the organism looks like a short, plump rod. This form is called a *coccobacillus* because of its tendency to resemble a coccus even though it is actually a bacillus. In general, whenever you deal with a pure culture of a coccobacillus, there will be a *few* clear-cut examples of genuine rod forms in every field of view. When you see some nearly spherical forms that are peculiarly unlike typical cocci and you also see a few definite rods scattered about, you can assume to be dealing with a coccobacillus (if your culture is pure). It takes practice and concentration on the field of view to achieve skill in this kind of identification. Look at *individual organisms,* not whole fields of organisms. Since you will be given other tasks of this type in future experiments, it is good to firm up your technique on this point.

EXPERIMENT 2-15

Cytological Characterization of an Unknown Bacterium

MATERIALS

1. Unknown organism, assigned by number.

PROCEDURE: In this experiment you are assigned a 24-hour culture that is presumed to be pure. The culture is identified only by a number. Record this number in your notebook as soon as you receive the culture. Perform the following staining procedures on your unknown organism (or those stains assigned by your instructor): (1) fat inclusions, (2) metachromatic granules, (3) Gram stain, (4) acid-fast stain, (5) spore stain, (6) flagella stain, (7) capsule demonstration, and (8) motility (either by means of a wet mount or a hanging drop preparation). When you have concluded your study, you will be asked to turn in a record of your results. You will be informed as to the correctness of these results. It is important for you to remember that certain of these procedures are based upon the age of the culture, so take this into consideration when you plan your work. In general, it is best to do any of these procedures with a 24-hour culture, although it is problably better to do the spore stain on a 48-hour culture.

CHAPTER 3

NUTRITION OF MICROORGANISMS

BASIC GROWTH REQUIREMENTS

One of the most important facets in a study of living things is the science of nutrition. We commonly think of nutrition in terms of ourselves and our requirements for particular types of food, often not giving much thought to the idea that *all* living things are subject to nutritional needs as well. In the final analysis, nutrition deals with the kinds of food that an organism must have in its diet in order to survive. Since foods are actually combinations of various chemical substances, a proper approach to the science of nutrition puts it on a purely chemical level. Most of us are aware that we need a certain amount of protein in our diet to survive. However we don't often think of the chemical reasons behind this requirement.

All living things require some substance in their diet that contains nitrogen in a form *that they can use*. Human requirement of nitrogen is met by proteins or the amino acids from which they are formed. Looking at this in another way, one can see that proteins and amino acids provide man with usable nitrogen so his body cells can produce cellular protein. If man were to attempt to survive on a diet that was free of all nitrogen except for an inorganic salt such as potassium nitrate (KNO_3), man would not survive very long, because the nitrogen would be inadequate. Potassium nitrate contains nitrogen, but our cells lack the necessary enzymes for converting this form of nitrogen into protein.

Microorganisms also require a usable source of nitrogen in their diet. Many organisms are capable of using protein and amino acids as a source of nitrogen, but many organisms do *not* require such elaborate sources of this element. Inorganic nitrogen compounds, such as potassium nitrate and ammonium phosphate, can serve as the *only source* of nitrogen for various types of microorganisms. From these compounds, certain microorganisms can create the various amino acids and ultimately microbial protein. In the study of microbial nutrition, it is therefore common practice to prepare a diet for the test organism that is known to be complete except for available nitrogen. Various nitrogen compounds are then added singly to determine if growth can take place. In such an experiment, one need only observe the presence of *actively growing cells* to confirm that the organism is capable of utilizing this source of nitrogen. If no growth takes place, one can assume that the diet was inadequate with respect to nitrogen.

Another familiar dietary material is carbohydrate (sugars, starch, etc.). Carbohydrates generally supply readily available energy or fuel upon which the living organism depends. The source of energy for microorganisms is variable; some are able to utilize most types of carbohydrates while other microorganisms utilize only a few carbohydrates, if any. Still other microorganisms are capable of deriving their source of energy from purely inorganic substances, such as sodium carbonate. If man were forced to live on a diet in which sodium carbonate was the only available energy source, he would lose weight rapidly. In general, one finds that higher animals require very complex organic sources of energy. Many microorganisms also fall into this category, but again there are many that are far less particular in their requirements. Experiments concerned with the determination of usable sources of energy are usually devised so that the only source of carbon is a known substance, and one observes for growth, or no growth, as in the previous experiment with nitrogen. All other nutriments in such an experiment would be known to be acceptable to the organism in question.

Ideally all the various media (diets) employed in microbiology would be of the so-called synthetic variety. A *synthetic medium* is one whose exact structural chemical formula is known. By contrast, *most* microbiological media contain substances whose chemical formulas are unknown. Such substances as beef extract, various proteins, peptones, and yeast extract all represent a heterogeneous mixture of compounds of unknown chemical composition and structure. When proteins are subjected to acid or enzymatic breakdown, various-sized intermediate products accumulate which are just linear chains of amino acids of varying lengths. They have been generally called, in descending order by size, *proteoses, peptones,* and *peptides.* In any such preparation there will also be free amino acids. Any of these preparations can be standardized to a reasonably uniform product by careful control of the starting material and the conditions of hydrolysis, but there is no way of knowing the precise chemical formulations of the hydrolysate. Whenever such substances are added to a medium, it can no longer be synthetic, even though these substances are mixed as specific ingredients. Much can be learned regarding the nature of microbial nutrition through the use of synthetic media. In the following experiment you will utilize simple media to demonstrate that not all microorganisms have the same nutritional requirements.

EXPERIMENT 3-1

Determination of Minimal Growth Requirements

MATERIALS

1. Talls containing the following types of media (1 each):
   a. Agar only (washed or purified before dissolving).
   b. Agar + minerals (agar washed as above).
   c. Agar + minerals + organic carbon source (sugar).
   d. Agar + minerals + sugar + organic nitrogen source.
2. Sterile Petri dishes (4).
3. Culture of <u>Escherichia coli</u>, <u>Streptococcus lactis</u>,* and <u>Saccharomyces cerevisiae</u>.

PROCEDURE: Melt a tall of each of the different kinds of media and allow to cool to 45°C. Pour each cooled tall into a separate sterile Petri dish, allow to harden, and label. Divide each Petri dish into four sectors by writing on the bottom with a wax pencil. Inoculate corresponding sectors of each plate (these should be numbered) with one of the three organisms, leaving the fourth sector of each plate uninoculated. Invert the plates, incubate for 48 hours and observe for growth in each of the inoculated sectors. The plate labeled "agar only" contains nothing more than distilled water and plain agar. The minerals in the other three media include ammonium phosphate ($NH_4H_2PO_4$), potassium chloride (KCl) and magnesium sulfate ($MgSO_4$). The organic carbon source used in the last two media is glucose sugar, and the organic nitrogen source used is a bacteriological peptone. In light of your observations, what interpretations can you make regarding the ability of these different media to support microbial growth?

The student should examine the various media formulations and analyze the contents of each medium as to source of nitrogen, energy, and accessory growth factors. The appendix of this book lists some of the formulas in use, but a much more extensive listing can be obtained from commercial sources of bacteriological media (e.g., Difco, and Baltimore Biological). A good understanding of what each ingredient is used for will give the student much more insight into many of the laboratory experiments to come.

ADAPTATIONS OF MEDIA TO STUDIES OF MICROORGANISMS

A knowledge of the nutrition of a wide variety of microorganisms can be put to a more practical use in the microbiology laboratory. Many times it is desirable to isolate and identify a single type of organism from a very

---

**Streptococcus lactis* is a round chain-forming bacterium which is most frequently associated with souring milk or milk products. Certain strains of this organism are commonly used in the preparation of various cheeses and other commercial products.

populous mixture of different kinds of organisms. At first glance, this would appear to be something like the old "needle-in-a-haystack" situation. However, by utilizing knowlege of microbial nutrition, it is possible to narrow the search rapidly. The usual procedure is to employ a *selective medium.* A selective medium is one that has been formulated chemically so as to allow growth of the species you are seeking to isolate, but also to suppress the growth of most other competing species. In this way, the inoculation of a selective medium with a mixture of many different kinds of microorganisms will rapidly narrow the field, simply because most of the undesired microorganisms will not grow.

Another technique used in this kind of study is to take advantage of known chemical reactions resulting from microbial growth on a particular kind of medium. A frequent change resulting from microbial growth is the alteration of the existing pH of the medium. Here it is possible to formulate the medium so that it can be determined at a glance whether a certain colony is bringing about a specific chemical change or not. Such a medium is usually called a *differential medium.* Ordinarily such a chemical change is made evident by the use of colored indicators. If an indicator is incorporated into a medium at one color (i.e., pH) and the growth from the desired species changes the pH around the developing colonies, a different color will appear.

These are but a few of the concepts that are basic to the study of microbial metabolism. Ultimately, the basis for the various growth requirements rests upon the chemical capabilities of the cells in question. Subsequent experiments are, therefore, designed to demonstrate some of the variations in chemical capabilities that exist among the different species of bacteria. Keep in mind that the concepts illustrated here apply to all living things and are not applicable to microorganisms alone.

EXPERIMENT 3-2

Selective and Differential Media

MATERIALS

1. Nutrient agar for plating.
2. Sodium chloride agar tall (1).
3. Phenol red agar tall (1).
4. Cultures of Staphylococcus epidermidis and Alcaligenes faecalis.*
5. Petri dishes (3).

PROCEDURE: Pour one sterile Petri plate with nutrient agar, another with sodium chloride agar, and a third with phenol red agar. Divide each plate into three sectors, labeling one sector Alcaligenes

---

Alcaligenes faecalis is a very small rod-shaped bacterium that is commonly found in the intestinal tract of man, but it is also widely distributed in decomposing matter in nature and is considered to be nonpathogenic.

faecalis, another sector <u>Staphylococcus epidermidis</u>, and leaving the third sector unlabelled. Streak each culture in the appropriate sectors on each of the Petri plates with your inoculating loop. Use a very small quantity of growth and attempt to get isolation if possible. Invert the Petri plates and incubate for 24 to 48 hours. The phenol red agar medium contains glucose as a fermentable carbohydrate. Many organisms attack this substance and convert it into acid end products. From your observations of the results obtained, which of the three media used do you consider to be selective and which do you consider to be <u>differential</u>? Explain your answers.

The next experiment utilizes synthetic media to ascertain various growth requirements. Each of the three media employed are chemically defined (see Appendix). You are to inoculate two organisms in each medium and observe for growth after a suitable incubation period. If growth occurs, you can assume that minimal growth requirements have been met. To understand the results of this experiment you must first familiarize yourself with the ingredients of each medium. The only nitrogen source in uric acid broth is uric acid, while the nitrogen source in the other two is ammonium phosphate. The carbon sources are glucose in both the uric acid and ammonium phosphate media, and sodium citrate in the sodium citrate medium.

EXPERIMENT 3-3

Utilization of Unusual Sources of Nitrogen and Carbon by Microorganisms

MATERIALS

1. Cultures of <u>Escherichia coli</u> and <u>Aerobacter aerogenes</u>.
2. Tubes of uric acid broth, sodium citrate broth (or Simmons citrate agar slants), and ammonium phosphate broth (2 each).

PROCEDURE: Inoculate tubes of each of the three media with each organism and incubate for 48 hours. After removing these tubes from the incubator, observe for growth within the tubes as indicated by turbidity or indicator color change if Simmons citrate agar is used. Be sure to agitate the fluids in the tubes in case any growth has settled to the bottom. Make a Gram stain on every tube showing turbidity to confirm the presence of the organism that you inoculated. What conclusions do you draw regarding growth requirements of the two organisms?

CHAPTER 4

METABOLISM OF MICROORGANISMS

EXOENZYMES OF MICROORGANISMS

One of the most spectacular aspects of microbial activity concerns the metabolic capabilities of microbial cells. Metabolism can be defined as "the sum total of all chemical activities of the living system." Microorganisms live at an astounding pace. When we compare bacteria with ourselves, we can see certain obvious features that illuminate this point. We tend to think of a generation as consisting of 20 years, whereas in many bacteria it is only 20 minutes. Our metabolic rate is many thousands of times slower than that of most of the bacteria. It has been estimated that the protists account for about 90% of all the carbon dioxide production on the earth, i.e., they collectively respire to this extent and animals and plants share responsibility for the remaining 10 percent. What they lack in size, therefore, they make up by an extremely high rate of chemical activity.

In addition to the extremely rapid rate at which they decompose raw materials, the many species of microorganisms collectively contain a great variety of enzymes capable of attacking virtually every type of organic substance known. Enzymes are proteins, manufactured by cells, that catalyze specific chemical reactions.

> Catalysis is a process whereby the rate of a given chemical reaction is changed. In living cells, catalysis is ordinarily an acceleration or speeding up. Therefore, enzymes are biological catalysts that accelerate specific chemical reactions within the cell. This function is of fundamental importance to life itself, since the "pace of living" would be fantastically slow without the catalytic action of enzymes. Reactions that would normally take several hundred years to complete are reduced to a split-second duration by the action of appropriate enzymes.

The greater the diversity of enzymes an organism possesses, the more different raw materials it is capable of using metabolically. Some microorganisms have so many enzymes that they are capable of living upon nothing more than simple inorganic material; however, many other microorganisms require virtually all food to be pre-formed in a complex organic state similar to foods in the diet of man and higher animals, for example, amino acids and glucose.

In a laboratory study of microbial metabolism, attention is first focused upon the location of action of the various enzymes. By far the greatest number of enzymes are found within the individual cell, and these enzymes are capable of acting on substances only after the substances have entered the cell. Another group of enzymes, fewer in number, are found outside the cell and these are capable of acting independently on substances found in the environment of the organism. Enzymes that act outside the cell are called *exoenzymes;* enzymes that act inside the cell are called *endoenzymes.* Many complex organic substances in the environment of microorganisms are too large to enter the cell. The principal function of the exoenzyme is to reduce the molecular size of many of the complex substances in the environment until they can diffuse into the microbial cell where they can be acted upon by the endoenzymes. Exoenzymes are primarily *hydrolytic* enzymes; that is, an enzyme that ultimately acts to reduce the size of a molecule (usually changing the substance from a colloidal state to a crystalloid state) with a corresponding addition of water to the resulting products.

> The terms "colloid," "cyrstalloid," and "suspensoid" are used in an artificial but convenient system for classifying the size of particles dissolved in a solution. The usual rule of thumb is to classify *crystalloids* as all particles, atomic or molecular, that are less than 1 millimicron in size; whereas *colloids* are particles that fall into the 1 to 100 millimicron size range; and finally *suspensoids* are those particles whose diameter is greater than 100 millimicrons. They can be further differentiated on the basis of their gravitational properties. Under the influence of the most powerful centrifuging, crystalloids cannot be separated from the solvent in which they are dissolved. At ultracentrifugal force, colloids can be forced out of solution (ultracentrifugal force is usually regarded as that attained at speeds above 50,000 revolutions per minute). Suspensoids are particles that settle out on standing and therefore require no centrifugal force other than normal gravity to separate them from the solute in which they are dispersed. Examples of crystalloids (so-called because most such particles form crystals when dried) are: common table salt, sugar, and amino acids. Colloidal substances include compounds of high molecular weight like proteins, polysaccharides, nucleic acids, and many of the higher fats. Suspensoids are represented by cellular particles, such as bacteria and red blood cells, and grains of sand. In general, one can say that biological membranes are usually incapable of permitting entry of particles greater than crystalloid size, although there are some exceptions to this.

In subsequent experiments, you will observe bacterial exoenzymes in action. These are often easily demonstrated by certain visible changes that occur in the medium in which the colony is growing. As the exoenzymes are produced, they diffuse into the medium surrounding the mass of growth. Once the enzymes are free of the cells, they begin acting upon materials in the agar (provided the material is appropriate for the enzyme in question, since enzymes are usually specific in their action). The enzymes therefore function independently of the cell from which they come. By "seeding" a medium with a given colloidal substance and subsequently examining the agar surrounding the bacterial growth for its disappearance, the student can experimentally demonstrate the action of various exoenzymes.

EXPERIMENT 4-1

The Hydrolysis of Starch

MATERIALS

1. Cultures of Bacillus subtilis and Escherichia coli.
2. Starch agar tall (1).
3. Gram's iodine solution (diluted with water 1:2).
4. Sterile Petri dish (1).

PROCEDURE: Melt the starch agar tall and pour into the sterile Petri dish. Starch agar is nutrient agar to which a small amount of starch has been added as the only available carbohydrate. One of the organisms to be planted on the agar surface produces an enzyme capable of hydrolyzing starch. This enzyme must be an exoenzyme, because the starch molecule is colloidal in nature. Millions of bacteria releasing billions of starch-hydrolyzing particles into the medium have the overall effect of destroying the starch in the immediate vicinity of microbial growth, hydrolyzing it ultimately to maltose or glucose (crystalloids). After the agar has hardened, trace a single streak at one edge of the Petri plate with E. coli and on the other edge with B. subtilis, as in the following diagram (one on each side, well separated):

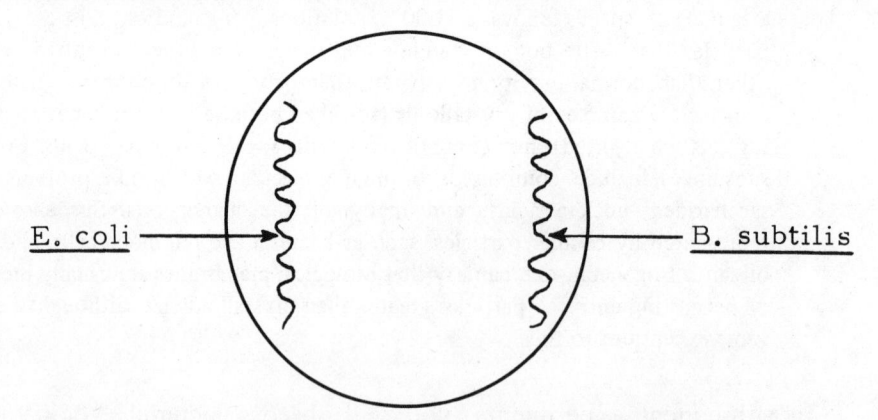

Label the plate and incubate at 37°C. for 48 hours. After incubation is complete, flood the surface of the plate with dilute Gram's iodine solution. As you know, whenever iodine and starch are mixed together, a dark-blue color results. This reaction is quite specific, and the end products of starch hydrolysis do not give this reaction.

Examine both parts of the plate and record your observations regarding the ability of these organisms to produce a starch-hydrolyzing enzyme. The two names ordinarily given to starch-hydrolyzing enzymes are "diastase" or "amylase." Either name is correct.

The following experiment is similar to the one just completed, but in this case the colloidal substance to be hydrolyzed is casein — the principal protein of milk. A small amount of milk is dispersed in nutrient agar to provide casein for the organisms planted on the agar. The agar will acquire a distinct turbidity, which would not ordinarily be seen in a regular nutrient agar plate. This turbidity results because the casein molecules are colloidal, and most colloids do not transmit light readily. Rays of light passing into colloidal solutions are often deflected and do not pass through. This is why you cannot see through a glass of milk, whereas a solution of sugar or salt is perfectly clear. This turbidity is a function of the particle size of the colloidal molecules. While growing on such an agar, some microorganisms are capable of releasing exoenzymes that hydrolyze protein. The hydrolysis of casein is not different from that of any other typical protein and proceeds from the protein molecule through a series of intermediate-sized molecules called proteoses, peptones, peptides, finally resulting in the crystalloid-sized amino acids of the protein. Thus, when casein is hydrolyzed, it loses the optical property of reflecting light, and instead, transmits it. The change observed on a milk agar plate is the clearing of the agar immediately around the colony of organisms that produce the *caseinase*. In the previous experiment the colloidal starch was hydrolyzed to the crystalloid sugars, maltose and/or glucose. Once the crystalloid is produced from the larger molecule, it is free to diffuse into the cells growing nearby. In this way the organisms procure a supply of nutriment that would be unavailable if it were not for the enzymes. Another variation of this same experiment can be performed in a test tube in which bacteria are incubated in previously sterilized milk.

EXPERIMENT 4-2

Hydrolysis of Casein

MATERIALS

1. Cultures of Bacillus subtilis and Escherichia coli.
2. Nutrient agar tall (1).
3. Tube of sterile, fresh, skim milk (1).*
4. 1 ml. pipets (2).
5. Petri dish (1).

---

*The milk is sterilized by being boiled for 20 minutes on 3 successive days. Do not autoclave.

PROCEDURE: Melt a nutrient agar tall and allow it to cool to about 45°C. Aseptically pipet 1 ml. of sterile fresh skim milk into the tall. To prevent parts of the agar from solidifying prematurely, the sterile skim milk should be kept in the 45°C. water bath along with the tall prior to pipetting. After the milk has been added to the agar, close the agar cap and thoroughly mix the two solutions by rapidly rotating between the palms of your hands. After complete mixture has been obtained, pour the <u>milk agar</u> into a sterile Petri dish and allow it to harden. Make a single streak along the edge of the Petri plate with the culture of <u>B. subtilis</u>. Prepare a second streak in the same fashion with the culture of <u>E. coli</u> on the other side of the same plate, and incubate the plate for two days at 37°C. Observe your Petri plate for evidence of hydrolysis of casein, and make a sketch of your Petri dish in the laboratory report.

The next demonstration of exoenzyme action involves the enzyme *gelatinase.* As the name indicates, the substrate for this enzyme is the protein gelatin. The medium employed in this experiment will support the growth of most organisms, but the solid character of the medium depends on the gelatin remaining in the *gel* state. Some colloids can exist in either a sol state or a gel state according to conditions at the time. Ordinary gelatin remains in the gel state as long as the temperature is less than approximately 25°C. The detection of the gel state is made by simply observing the familiar gelatin-like characteristics. When the temperature goes above 25°C., gelatin goes into the *sol,* or liquid, state. It will remain a sol as long as the temperature is above 25°C. Reducing the temperature below 25°C. causes gelatin to return to the gel state. Although this reversible action can be continued almost indefinitely, the ability of gelatin to do this depends on its remaining a colloid. Many microorganisms produce exoenzymes that are capable of hydrolyzing gelatin. If such an organism is planted in a tube of nutrient gelatin, the gelatin molecules in the immediate vicinity gradually become hydrolyzed from the colloidal state to the crystalloid (amino acid) state. As a result of this transition, gelatin is no longer capable of exhibiting gel characteristics even when the temperature is lowered. Therefore, *true hydrolysis* of gelatin is indicated by its inability to become solid even after lower temperatures are maintained for a continuous period.

In a tube of inoculated nutrient gelatin any gelatinase production is indicated first by a liquifying of the gelatin along the line of the stab. This may occur at the surface only or along the entire line of stab, and it may proceed in peculiar patterns. The outermost portion of gelatin (i.e., that farthest from the line of stab) will hydrolyze last, if at all. Repeated observations have shown that certain species of bacteria hydrolyze, or liquify, gelatin in a particular fashion. This pattern formed by the liquified gelatin is often useful in identification because it is a repeatable phenomenon. On page 26 in Chapter 1, are illustrated some typical patterns of growth related to the gelatin stab technique. There are two possibilities — an organism will hydrolyze gelatin or it will not. If the organism does not hydrolyze gelatin, it will generally grow within the medium in one of four different patterns. On page 26, four typical patterns are illustrated, entitled *Line of Puncture* (non-liquifying). These patterns are observed with microorganisms that *do not* liquify gelatin. Any organisms in this experiment that do not liquify gelatin should be categorized according to one of these four patterns. On the other hand, if an organism *does* liquify or hydrolyze gelatin, five additional patterns can be observed by inverting the tube and shaking it so that the liquified gelatin flows out, leaving a "hollowed out" area within the gelatin. Five different patterns have been named and are diagramed on the same page for your reference. Any organisms used in this experiment that liquify gelatin should also be categorized according to one of the five other possibilities. This technique is quite useful in the identification of microorganisms.

EXPERIMENT 4-3

The Hydrolysis of Gelatin

MATERIALS

1. Cultures of Proteus vulgaris, Bacillus subtilis, and Escherichia coli.
2. Tubes (shorts) of sterile nutrient gelatin (3).
3. Talls of 0.4% nutrient gelatin agar (3).
4. Saturated ammonium sulfate solution.
5. Petri dishes (3).

PROCEDURE: Inoculate each of the organisms provided into a separate tube of nutrient gelatin with your inoculating needle by stabbing the inoculum to the bottom of the tube in the center of the gelatin and withdrawing along the line of stab. Incubate all tubes from two to seven days in your drawer, making daily observations after the second day. When you use this method, you must always incubate at room temperature, because the gelatin will melt if you put it in the incubator, thereby destroying any patterns of hydrolysis. After incubation, record your observations in your laboratory notebook, according to instructions given in the discussion previous to this experiment.

Another way to illustrate the same phenomenon utilizes a Petri plate (Fraser's method), but does not give pattern type information. Melt and cool three talls of 0.4% nutrient gelatin agar, and pour them into three sterile Petri dishes. After the agar has hardened, let the dish set for about 30 minutes so that the surface will become relatively dry. Streak one edge of the first dish with P. vulgaris, one edge of the second dish with B. subtilis, and one edge of the third dish with E. coli. Incubate all plates in the 37°C. incubator for two days. In this medium, gelatin is incorporated with nutrient agar. Any loss of gel state by the gelatin does not show a change to liquid by the medium, because the agar maintains the gel state of the entire medium. However, if the organisms grown on the surface of this medium have hydrolyzed the contained gelatin, it is easily determined by using a general protein-precipitating agent, in this case saturated ammonium sulfate. Flood each plate with about 10 ml. of the saturated ammonium sulfate solution and let the plate stand for 30 minutes. Wherever intact (unhydrolyzed) gelatin remains, a turbidity results due to the precipitation of this protein by ammonium sulfate. On the other hand, if the gelatin has been hydrolyzed, turbidity does not appear since the hydrolyzed gelatin does not precipitate. Agar is not affected by ammonium sulfate. Record results on all tests according to instructions given.

Probably the principal function of exoenzymes to microorganisms is the procurement of usable food. However, another aspect of exoenzyme action often overlooked is that many bacteria produce enzymes that serve as an "aggressive mechanism" or "weapon" in the give-and-take of the infectious process. Some of these enzymes are protein particles that allow the enzyme-producing organisms to establish themselves and grow in the tissues of man and other higher forms, despite the attempts of the host to prevent it. In the subsequent experiment some activities of these exoenzymes are illustrated. Earlier experiments illustrated the ability of selected organisms to hydrolyze various colloidal substances normally considered food items, such as milk, protein, and starch. In the following experiments, human or animal cells, or the products of these cells, rather than food molecules, become the target of destructive enzyme activity.

The next experiment customarily appears in the section on medical microbiology, but many times the student loses sight of the fact that this experiment should properly be included in a study of exoenzymes. The medium used in this experiment is nutrient agar to which whole blood has been added to a final concentration of 5%. This extremely rich medium supports the growth of many of the more nutritionally fastidious microorganisms. Many of these organisms produce exoenzymes that have a destructive effect on red blood cells. Such enzymes are referred to as *hemolysins*. Some microorganisms elaborate *beta* hemolysins, which destroy and decolorize hemoglobin from the red cell. This can be observed on a blood agar plate by noting a clear zone around the colony. Other microorganisms produce hemolysins of the *alpha* type, which do not decolorize the hemoglobin, but rather change it to a greenish color.

EXPERIMENT 4-4

Hemolysin Production

MATERIALS

1. Cultures of Staphylococcus aureus* (or Bacillus subtilis) and Diplococcus pneumoniae** (or Streptococcus lactis).
2. Blood agar plate (1).

PROCEDURE: Divide a blood agar plate into two equal halves with a wax pencil and streak for isolation, using half of the plate for each organism. If the pathogenic Staphylococcus aureus and Diplococcus pneumoniae are used in this experiment, be very careful in all phases of handling them. If you use proper sterile technique you should have no difficulties with them, but sloppy work can result in problems. If you should spill any of either culture, be sure that you flood the area immediately with disinfectant and allow it to act for 30 minutes or more. Incubate for 48 hours at 37°C., and observe for hemolysis. It is helpful to use a lighted colony counter in order to observe this phenomenon. Record your results.

Normal human or rabbit plasma, if allowed to stand, clots in the same manner as whole blood. This can be prevented by the addition of some sodium citrate (a general anticoagulant), after which the plasma remains fluid. Plasma thus treated can be used to illustrate another type of exoenzyme, coagulase, which is elaborated by the pathogenic organism *Staphylococcus aureus*. This substance helps the *Staph. aureus* establish itself in host tissue. How coagulase accomplishes this is not known, but it is known that all pathogenic staphylococci produce it. Coagulase reverses the action of the anticoagulant (sodium citrate, in this experiment). When this plasma is inoculated with pathogenic *Staph. aureus* it will clot, just as if no citrate had been added.

EXPERIMENT 4-5

Coagulase Production (Demonstration)

MATERIALS

1. Culture of Staphylococcus aureus.
2. Samples of citrated human or rabbit plasma (Commercially available in dehydrated form).

---

*Staphylococcus aureus is a round bacterium that is normally found in the mucous membranes and skin of man. It is a pathogenic organism responsible for a wide variety of infections of man.
**Diplococcus pneumoniae is an oval-shaped bacterium that produces a capsule and is a normal inhabitant of the human throat and oval cavities. It is also a very common cause of lobar pneumonia in man. It is somewhat difficult to grow in the laboratory.

PROCEDURE: This experiment will be demonstrated for you, but you are asked to observe its progress and report on it as if you had performed it yourself. The performance of the coagulase test is very simple. A loopful of surface growth from an agar slant of Staph. aureus is introduced into a small test tube (10 mm. × 75 mm.) of citrated plasma and incubated for 2 or 3 hours. Every 30 min. the plasma is observed for clot formation. This is best done by tilting the tube to see if the plasma is still fluid. When the tilting action does not cause the plasma to flow down the tube as it did at the beginning of the experiment, the end point of the experiment has been reached. An uninoculated control tube should be included in this experiment for comparative purposes.

ENDOENZYMES OF MICROORGANISMS

The enzymes of microorganisms and other cells are found in both intracellular and extracellular locations. You have considered some of the exoenzymes manufactured by various microorganisms; now we turn our attention to those enzymes that are found within the cell and are generally referred to as endoenzymes. The microbial cell is somewhat like a chemical factory, and the factors within the cell that bring about the multitude of chemical changes are the various endoenzymes. This chemical activity is essentially the conversion of entering food molecules into various chemical products that can be further utilized by the cell (a) to create more cell substance or (b) to degrade incoming molecules in order to extract energy from them. All living cells must carry on at least two basic activities — the synthesis of new cellular material, and the constant removal of energy from compounds by the enzymes. The former reactions are generally referred to collectively as the *synthetic* reactions, and the latter collectively as the *respiratory* reactions. These activities ultimately produce an accumulation of substances that are no longer usable by the cell. These substances are the *waste products* of metabolism. The waste products continually diffuse out of the cell and into the environment, just as nutrients are constantly being incorporated into the cell to undergo this transformation.

The inflow and outflow of materials in this fashion are common to all living cells, but different types of cells differ from one another depending on what they do with the incoming nutrients, and therefore they also differ in the types of waste products they release. These differences are the result of the action of *different types of endoenzymes.* Some organisms can live in exceedingly simple media because they have within them the proper enzymes necessary to carry out all synthetic and respiratory functions, starting with the simple materials in the medium. Organisms requiring complex organic media generally lack many enzymes and therefore must have many of the nutrient substances *pre-formed* before the organisms can make use of them.

In the experiments to follow, you will consider some of the variations in enzyme content of different microorganisms. You will by no means make a comprehensive survey of the field but rather a sampling to illustrate the principles involved.

An analysis of the end or waste products of respiratory activities can be utilized to determine the presence of certain groups of enzymes within different types of cells. Some microorganisms exhibit a very simple respiratory pattern in which only a single waste product is released. In such organisms, the entire sequence of chemical reactions leads to but one end product. An analysis for the presence of this particular end product, then, is of some assistance in the identification of the organism. On the other hand, many bacteria, because they contain a wide variety of enzymes, will give off many different waste products from the *same* starting material.

Respiration can be carried out under both aerobic and anaerobic conditions, depending upon the type of organism. Most living organisms are capable of carrying out aerobic respiration, which ordinarily completely oxidizes the beginning product to *carbon dioxide* and *water*. Most organisms do not differ markedly as to the method of aerobic respiration. It is under anaerobic conditions that one finds major differences in what is done with the starting product by the enzymes of the microorganisms. In the first of the experiments to follow, you will observe the fermentation of the starting material by a microorganism that is capable of fermenting many sugars, with the formation of *alcohol* and *carbon dioxide* as end products. This reaction is basic to the fermentation industry. All alcoholic beverages produced in this country are manufactured by the action of such microorganisms. In another experiment, you will see that fermentable sugars are converted into *lactic acid* as the only end product of fermentation. Both processes are quite similar. Starting with a typical sugar such as *glucose* (also called dextrose), most organisms will enzymatically convert this substance to *pyruvic acid,* and from this point the various end products of fermentation result. Pyruvic acid is a "crossroads" in the respiratory process, since most organisms first convert fermentable sugars to this product. From pyruvic acid, however, to the waste product (or products), the respiratory patterns vary widely. The general reaction may be pictured as follows:

```
        glucose
           |
           v
      pyruvic acid
         / | \
        v  v  v
   various end products
```

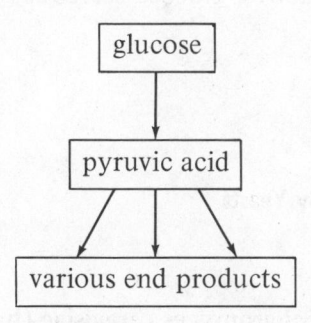

In yeast fermentation, pyruvic acid is converted enzymatically into acetaldehyde (called a metabolic intermediate) and this is finally reduced with hydrogen to *alcohol,* with *carbon dioxide* liberated, as indicated in the following equation:

$$\text{glucose} \longrightarrow \text{pyruvic acid} \longrightarrow \text{acetaldehyde} \longrightarrow \text{ethyl alcohol}$$

In lactic-acid-forming bacteria, the chemistry is somewhat simpler. After pyruvic acid is formed, it is simply reduced to *lactic acid,* which is the end product, as illustrated in the following equation:

$$\text{glucose} \longrightarrow \text{pyruvic acid} \longrightarrow \text{lactic acid}$$

In both examples shown above, glucose serves as the source of hydrogen for the reduction reaction.

EXPERIMENT 4-6

Alcohol Production by Yeasts

MATERIALS

1. A culture of Saccharomyces cerevisiae (bakers' yeast).
2. Tube of malt extract broth (1).
3. Dried raisins.
4. Distilled water.
5. Clean test tube (not sterile) (2).

PROCEDURE: Aseptically inoculate a tube of malt extract broth with the culture provided and incubate in your drawer. Next put four or five dried raisins into a clean test tube. Add enough distilled water to cover the raisins, then use your glass rod to mash the raisins in the water so as to make a suspension of crushed raisins. Be sure not to push the bottom of the test tube out. Put this test tube in a water bath and bring the water bath to a boil. Keep the test tube in the water bath for about 5 to 10 min., occasionally stirring and crushing the fruit with your glass rod. Cap the tube and allow it to cool to room temperature. Inoculate this suspension with the culture provided as you did the malt extract. Repeat this procedure using a third test tube, but this time do not boil the extract and do not inoculate it with the culture provided. Thus this tube will contain a nonsterile suspension of mashed raisins. Incubate all tubes in your drawer. After a one week incubation, remove the cap from each tube, and note the characteristic odor of ethyl alcohol. A qualitative chemical test for ethyl alcohol could be performed if you wished more precise evidence of its presence, but ordinarily the odor is sufficient for detection. Note also in each of your test tubes the presence of gas bubbles, particularly when you agitate the tubes with a rotary motion. Record your observations.

EXPERIMENT 4-7

Lactic Acid Production by Bacteria

MATERIALS

1. Cultures of Lactobacillus casei* and Streptococcus lactis.
2. Tubes of sterile fresh skim milk (2).
3. Tubes of nonsterile skim milk (boiled 3 days for 20 minutes each day) (2).
4. Dilute hydrochloric acid solution (0.1 N).

PROCEDURE: Inoculate a tube of sterile skim milk with L. casei and another tube with Str. lactis, using aseptic precaution. Label a third tube of skim milk that has not been sterilized and do not inoculate it. Incubate all tubes in your drawer for about four days. At the end of the incubation period, examine all tubes for evidence of microbial change. When all tubes appear to be approximately the same in terms of curd formation, make a gram stain from each tube and also note the characteristic sharp fermentative odor present in each tube. After examining the gram stains under oil immersion, what conclusions do you draw regarding the presence in sour milk of

---

Lactobacillus casei is a normal inhabitant of milk and milk products and is often used as a cheese-producing organism in industry. It is a rod-shaped organism.

the various types of organisms? Make drawings from your stains, labeling each appropriately. The end product of this fermentation is simple lactic acid.

Because the pH within the tube was lowered, a physical change was brought about in the character of the protein normally present in milk. The protein in milk is predominately casein, and secondarily lactalbumin. When the pH is lowered to a certain point, these substances are made insoluble. You see evidence of this by the formation of a curd. Proof of the fact that an increase in acidity was the only necessary change to take place in the tubes can be illustrated very simply. Procure another tube of ordinary nonsterile skim milk and add a few drops of dilute hydrochloric acid to the tube and observe the results. Note the similarity between the contents of the second tube and the tubes that underwent microbial action. The lactic acid produced by the microorganisms normally present in milk is responsible for the normal souring of milk. It is through such a process that we obtain food products such as buttermilk and yogurt.

The term *fermentation* is ordinarily applied to the anaerobic degradation of carbohydrates, although other products besides carbohydrates can be fermented. In the experiment to follow, you will deal primarily with the anaerobic dissimilation of various carbohydrates by microorganisms. Ultimately the purpose of fermentation is to make energy available to a living cell, since carbohydrate molecules are quite rich in stored energy. Whether or not a given carbohydrate is fermented depends upon the enzymes contained by the organism. The *end products* of fermentation vary from one organism to another, also depending upon the enzymes contained by the organism. You can detect some of these end-products in an indirect fashion. For example, you have seen that the fermentation of sugar by yeast yields alcohol and carbon dioxide. The observation was made by smell and by seeing bubbles in the medium. Although neither of these procedures are considered scientifically critical, for practical purposes they are useful. In the experiment to follow on the fermentation of carbohydrates, you will use similar methods for detecting the end products of metabolism. In the fermentation of lactose to lactic acid by organisms in milk, you observed that proteins of milk coagulated. Acid production can also be determined by the use of indicators. In the following experiment, the tubes containing the media plus corresponding carbohydrates have an added pH indicator called *phenol red.* You have familiarized yourself with this indicator to a certain extent, and now you will see its practical use in microbiological media. In addition, Durham tubes will be used to collect any gas that might evolve as an end product.

Finally, it must be remembered that each fermentation broth contains only a *single specified carbohydrate.* If the organism included is capable of attacking that particular carbohydrate, certain changes will be observed after incubation. If the included organism is not capable of attacking the particular carbohydrate, *growth will occur* (due to the nutrient broth present), but no change in pH will be seen, with the possible exception of a change to a basic condition. Many times microorganisms select an alternative source of energy when the available carbohydrate is not usable. In other words, if they do not possess the proper enzyme to attack the particular carbohydrate, another source of energy must be found. In this case, amino acids may be utilized by a process that converts them into carbohydrate-like molecules, usually alpha-keto acids. This is done by removing the amino group (NH_2) from the amino acid and oxidizing the remaining portion by substituting an oxygen atom in place of the amino group. This produces the alpha-keto acid, which can be utilized in a fashion similar to that employed in obtaining energy from carbohydrates (i.e., the alpha-keto acid is then burned like a sugar). The ammonia that is evolved by the removal of the amino groups dissolves in the solution, and after enough microorganisms have performed this process, the medium usually becomes quite basic owing to the accumulation of ammonium hydroxide. A change in pH in the basic direction is indicated by a deepening of the red color of the phenol red, often to purplish-red.

We might summarize the ingredients present in the fermentation broth as being: (a) *nutrient broth,* which will support the growth of most microorganisms; (b) a *single chemically defined carbohydrate;* (c) a pH indicator; and (d) an inverted test tube for collecting any gases evolved. The recording of your results in this experiment should be designated as follows:

NC = No change in pH
A = Acid only
AG = Acid and gas
B = basic

Since there are many fermentable sugars and other related compounds, the ability to ferment a given substance offers an excellent aid to distinguishing one organism from another.

EXPERIMENT 4-8

Fermentation of Carbohydrates by Microorganisms

MATERIALS

1. Cultures of Escherichia coli, Alcaligenes faecalis, and Staphylococcus aureus.
2. Sucrose, lactose, and glucose fermentation broths in Durham tubes (3 each).

PROCEDURE: Inoculate a tube of each kind of fermentation broth with each of the organisms provided and incubate at 37°C. for 24 to 48 hours. Record what you observe in the various test tubes (as indicated in previous discussion) for both the 24- and 48-hour intervals. It is important to realize that these results can and sometimes do change with prolonged incubation. There is the possibility that acid fermentation will become basic on prolonged incubation, or that an apparent no change in the broth can become an acid change as a result of prolonged incubation. It therefore is emphasized that observations must be made daily for periods up to a week or so when you are attempting to identify organisms in this manner. Observe the tubes you have prepared for seven days; there is no necessity for recording anything beyond the 48-hour period unless a significant change takes place. Another precaution involves the use of phenol red. The pH at which this indicator changes color is about 6.8 or 6.9. Therefore only a small amount of acid production by the microorganisms changes the indicator to the acid color. Frequently there is a tendency on the part of some microorganisms to bring about an intermediate color change that might be described as somewhat orange. The rule of thumb to follow when using fermentation broth with phenol red is that if the full yellow color is not obtained, the fermentation is not to be considered positive.

Since fermentation takes place in the absence of molecular oxygen, the organism must obtain its energy from the substrate molecules without the use of this gas. Energy-releasing reactions are oxidation reactions, but oxidation may occur in a variety of ways. Since molecular oxygen is not available in the fermentation process, oxidation must be accomplished by some other means. Most oxidations that occur anaerobically are of the type in which *hydrogen is removed* from the substrate. One of the easiest ways to visualize removal of hydrogen from substrate compounds is based on the fact that hydrogen must be disposed of after it is removed. Hydrogen does not, as a rule, evolve molecularly, but rather it must be attached to some recipient compound. If no *hydrogen-acceptor compound* is available, then fermentation cannot occur. On the other hand, if a steady supply of some hydrogen acceptor is made available, then the organism can respire continuously until conditions change adversely. In aerobic respiration, *oxygen* serves the role of final hydrogen acceptor, but in the absence of oxygen other hydrogen acceptors can perform this task. One of the simplest ways of determining this effect is to utilize one of several colored compounds that lose normal color after reduction (hydrogen acceptance). Two substances of this type are the dyes *litmus* and *methylene blue*. Either compound is colored in the so-called oxidized or natural state, litmus being normally purple and methylene blue being normally blue. If these compounds are reduced by the addition of hydrogen, they lose their color. This offers a convenient indicator of hydrogen acceptor activity. Thus in the experiments to follow, you will observe after incubating microorganisms in the presence of these two dyes that the tubes of broth gradually lose their color, indicating that the fermenting organisms are using them as hydrogen acceptors. The reactions, greatly simplified, may be represented as follows:

$$\text{Litmus} \xrightarrow{+2H} \text{Litmus} \cdot H_2$$
$$\text{(purple)} \qquad\qquad \text{(colorless)}$$

$$\text{Methylene blue} \xrightarrow{+2H} \text{Methylene blue} \cdot H_2$$
$$\text{(blue)} \qquad\qquad\qquad \text{(colorless)}$$

You will probably note the slight ring of blue at the top of some of your tubes showing otherwise complete reduction of the dyes in the experiments to follow. In effect, this ring measures the depth of oxygen penetration into the medium because, given a choice of oxygen or a reducible dye, the organisms will utilize the oxygen. Thus the dye at the surface remains in the oxidized state.

EXPERIMENT 4-9

Reduction of Methylene Blue

MATERIALS

1. A large broth flask (500 ml.) containing 250 ml. of a young culture of *Escherichia coli*.
2. Clean test tubes (3).
3. Sterile pipets.
4. Methylene blue (Loeffler's).
5. Sterile nutrient broth (15 ml.)
6. Several milk samples of varying ages (optional).

PROCEDURE: Set up three clean test tubes, and pipet mixtures of the bacterial culture and sterile nutrient broth so that the first tube contains 6 ml. broth culture and 3 ml. nutrient broth, the second 4.5 ml. nutrient broth and 4.5 ml. culture, and the third tube 6 ml. nutrient broth and 3 ml. broth culture. Mix each tube well, and add two drops of Loeffler's methylene blue to each tube with mixing. Put the tubes in the 37°C. incubator and observe at intervals of 10 min. for any changes that take place. Do not shake the tubes. Record the time needed for total decolorization of each tube.

This experiment illustrates the effect of hydrogenation on methylene blue. Naturally the more microorganisms present in a given tube, the more rapidly decolorization takes place. This fact previously was utilized to determine the quality of milk. The assumption was made that the more bacteria present per unit of volume in a given sample of milk, the poorer was the quality of the milk. The test was standardized so that milk was considered good if it would not decolorize within a certain period, while other grades of milk displayed faster decolorization times. As an optional addition, try the above technique on several samples of milk of different ages — that is, two, three, and four days since purchase.

In the next experiment, microorganisms will be inoculated into milk that has had a suitable concentration of litmus added to it. Although the primary purpose of this experiment is to demonstrate the reducibility of litmus by the respiratory activities of certain microorganisms, other secondary lessons may be learned from the choice of medium in which the organisms are grown. Milk, by itself, is capable of supporting the growth of many microorganisms. The color of the inoculated litmus milk is the purple of neutral or alkaline litmus. If the inoculated microorganisms successfully reduce the dye, the color changes from purple to colorless, and the normal color of the milk (white) is restored. Therefore, one of the daily observations to be performed in this experiment is to see if the litmus changes to the colorless state.

Due to the nature of the medium, the student will be able to make many other observations in this same experiment. Perhaps it would be well to list the ingredients of milk to aid in these observations. The principal ingredients of milk are the proteins *casein* and *lactalbumin,* the fermentable sugar *lactose* (also called milk sugar), and a variety of minerals and vitamins. Litmus is added as an *indicator* as well as a reducible dye. It turns pink in the acid condition and more deeply purple in the alkaline condition.

You have already observed certain other changes that can take place when microorganisms are grown in milk as a medium, and you should have noted that curd production can occur. Actually, two different kinds of curd may form. If the sugar lactose is fermented to lactic acid or any other organic acid, the pH will probably be lowered to the point where the milk forms a very hard curd. This is called an *acid curd.* To determine if the curd is of this type, tilt the tube to a horizontal position and see if it will flow. If the curd remains a compact mass, even upon inverting the tube, it is definitely an acid curd. The other kind of curd is called a *rennet* curd. Some organisms produce the enzyme *rennin,* which is a clotting enzyme. The curd formed by rennin is a soft one, similar in appearance to the curd formed when artificial rennin is used to make custard. This curd has a typical custard-like appearance and when the test tube is tilted, it will flow, but very slowly. The rennet curd typically forms in the absence of acid production; if acid were produced, the acid curd would definitely dominate.

If the microorganism inoculated into the tube of litmus milk is incapable of fermenting lactose, it may have to seek additional sources of energy from the proteins present. Some organisms possess enzymes capable of hydrolyzing casein and/or lactalbumin. Digestion of milk protein is usually called *peptonization,* which means the same as *proteolysis* or *hydrolysis* of protein. Peptonization of a protein frequently results in the release of large quantities of ammonia, and the pH of the medium gradually becomes alkaline. The final test for peptonization, however, is based on a change in the physical characteristics of the protein in the milk from the colloidal to the crystalloid state. Because colloidal substances are ordinarily nontransparent, you cannot see through a tube of ordinary milk. The milk protein deflects the light, making the milk opaque. On the other hand, a test tube solution of amino acids derived from the proteins of milk is clear; you would be able to see through it when it is held up to light. If you observe such a change taking place in the following experiment, hydrolysis of the protein into amino acids should also result in the clearing of the medium so that it becomes translucent. This kind of reaction ordinarily takes longer than some of the others, so continued observation is necessary.

Finally, another reaction that can usually be observed during the seven-day period is the production of gas. This, of course, requires the presence of an organism capable of fermenting the lactose into acid and gas. If such an organism is present, the gas may usually be detected by a "blowing apart," or separation, of the curd. Gentle agitation of the tube should also help to display gas bubbles. Throughout your observations in the following experiment make sure that you *do not shake* or unduly disturb the tube, because many of these observations may become obscured if you do.

EXPERIMENT 4-10

Reduction of Litmus Milk

MATERIALS

1. **Cultures of Escherichia coli, Bacillus subtilis, and Streptococcus lactis.**
2. **Tubes of sterile litmus milk (boiled for 20 minutes on 3 successive days) (3).**

PROCEDURE: Inoculate one tube of litmus milk with E. coli, another with B. subtilis, and a third with Str. lactis. Put all tubes in the 37°C. incubator, and observe daily through a seven day period. The following reactions should be included in your observations: (a) reduction of litmus, (b) formation of acid or rennet curd, (c) hydrolysis of protein, (d) gas production, and (e) any changes in pH according to the indicator.

Many organisms can reduce nitrate salts (such as potassium nitrate, KNO_3) to a nitrite salt counterpart (potassium nitrite, KNO_2) under anaerobic conditions. This is an example of a process frequently referred to as anaerobic respiration, in which the nitrate molecule acts as an electron acceptor in place of oxygen. This reduction of nitrate can easily be demonstrated experimentally by formulating a medium containing nutrient broth (in which included organisms grow with or without any other added materials), to which a very small amount of potassium nitrate is added. Since the included organisms will be growing in the depths of the broth, the conditions will largely be fermentative. Under these circumstances, any available hydrogen acceptors are utilized if the organisms possess the proper enzymes. Therefore, organisms with this capability will utilize the added potassium nitrate as a hydrogen acceptor. The resulting product of this reaction is potassium nitrite, which will accumulate and be present in the medium, although *not* there at the beginning of growth. Proof of this activity involves performing a chemical test on the medium that is specific for nitrite salts but is not reactive with nitrate salts or any other ingredient of the media. In this experiment, the added reagent develops a color in the presence of nitrite but does not do so in the presence of nitrate. Positive color formation is proof of the formation of nitrites.

EXPERIMENT 4-11

Reduction of Nitrates

MATERIALS

1. Cultures of Escherichia coli, Pseudomonas aeruginosa,* and Bacillus subtilis.
2. Sterile nitrate broth (3).
3. Sulfanilic acid solution.
4. Dimethyl-α-naphthylamine solution.
5. Nessler's reagent.
6. Sterile pipets (3).
7. Clean test tubes (2).

PROCEDURE: Inoculate a nitrate broth tube with E. coli, another with Ps. aeruginosa, and a third with B. subtilis. Incubate at 37°C. for four to six days. Make daily observations on your tubes and, starting with the second day, carry out the following procedure on each tube: Aseptically transfer 1 ml. of the medium in which the organisms have grown into a clean test tube. To this, add three drops of the sulfanilic acid reagent, followed by two drops of the dimethyl-α-napthylamine solution. If nitrites are present, the mixture of these two reagents with the medium will become pink or red — a positive test for nitrites. When a positive test for nitrites has been obtained, incubate the tubes for one more day, and then perform the qualitative test for ammonia by aseptically transferring 1 ml. of the medium to a clean test tube and adding several drops of Nessler's solution. If ammonia is present, a yellow-brown color develops in the medium. Record your results showing daily observations regarding nitrite and ammonia production. Ammonia is a reduction product formed after the nitrite stage that some organisms are capable of producing after prolonged incubation.

Another activity of this type that actually falls into the category of fermentation of proteins involves the production of hydrogen sulfide. Some amino acids of most proteins contain sulfur. Some organisms possess the

---

Pseudomonas aeruginosa is widely distributed in soil and water in nature, particularly in polluted water. On occasion it can be pathogenic for man, and when this happens it tends to cause a serious infection. It is a small rod-shaped organism.

enzyme that splits off the sulfur atoms, which are subsequently reduced by the addition of hydrogen (from substrate molecules) to form *hydrogen sulfide*. Sulfur therefore serves as a hydrogen acceptor in this case. This reaction, although involving amino acids, does not differ in theory from those described earlier. To detect the ability of an organism to form hydrogen sulfide, one must grow the organism in the presence of sulfur-containing amino acids. The amino acid that is primarily utilized for this purpose is *cysteine*. The medium chosen for the next experiment contains large quantities of cysteine. Since hydrogen sulfide is a gas, a specific chemical test within the growth medium will be used to detect the hydrogen sulfide as it is formed. Such a medium contains an additive, either an iron or a lead salt, that reacts very readily with hydrogen sulfide to form the insoluble iron or lead sulfide (both are dark in color), as illustrated by the following equations:

$$H_2S + FeSO_4 \longrightarrow FeS + H_2SO_4$$
$$\text{(black)}$$

$$H_2S + PbSO_4 \longrightarrow PbS + H_2SO_4$$
$$\text{(dark brown)}$$

EXPERIMENT 4-12

The Production of Hydrogen Sulfide by Microorganisms

MATERIALS

1. Cultures of <u>Proteus vulgaris</u> and <u>Escherichia coli</u>.
2. Talls of sterile peptone-iron agar or triple sugar iron agar (TSI) (2).

PROCEDURE: Inoculate a tube of peptone-iron agar or TSI agar by <u>stab</u> inoculation with <u>P. vulgaris</u> and a second tube with <u>E. coli</u>. Incubate the tubes for 48 hours at 37°C., and observe for any blackening or darkening along the line of stab as evidence of hydrogen sulfide. It is occasionally necessary to make observations for four to six days for certain organisms, so before you decide that an organism does not form hydrogen sulfide, make sure you have observed for at least four days. In tabular form, show a daily record of your results.

To demonstrate that fermentation can take place with products other than carbohydrates, an additional experiment is included. Some organisms are capable of degrading the amino acid *tryptophan*, forming an end product of hydrolysis called *indole*, as shown in the following equation:

$$\text{Indole-CH}_2\text{-CH(NH}_2\text{)-COOH} + H_2O \longrightarrow \text{Indole-H} + CH_3\text{-CO-COOH} + NH_3$$
(Tryptophan) → (Indole) + (pyruvic acid) + (ammonia)

An experiment designed to detect this change utilizes a medium that is rich in tryptophan. This medium (tryptone broth) is inoculated and then a specific chemical test is performed to determine if indole has accumulated in the medium. The ability to form indole from tryptone is not displayed by all microorganisms and therefore serves as an aid in the identification of various microorganisms.

EXPERIMENT 4-13

Demonstration of the Production of Indole

MATERIALS

1. Cultures of Escherichia coli and Bacillus subtilis.
2. Tubes of sterile tryptone broth (2).
3. Kovac's solution.
4. 1 ml. pipets (2).
5. Clean test tubes (2).

PROCEDURE: Inoculate one tube of tryptone broth each with E. coli and B. subtilis, and incubate at 37°C. for approximately four days. After incubation is complete, aseptically transfer 1 ml. of the medium in which one of the organisms has grown to a clean test tube. Carefully add sufficient Kovac's solution (drop at a time) to form a layer on top of the medium approximately ¼ to ½ in. deep. Very gently agitate the tube with a rotary motion and allow to stand for several minutes. Do NOT agitate or shake so hard that the top layer is dispersed. It should remain intact on top of the medium. If indole is present in the medium, a cherry red color will appear in the top layer (Kovac's reagent). Repeat this procedure for the second organism. Record the results.

One of the most interesting experiments that can be performed in the area of bacterial respiration has given rise to a differential test known as the MR-VP test. The initials stand for *methyl red* and *Voges-Proskauer*, two different tests used for the same general purpose — to distinguish between *Escherichia coli* and *Aerobacter aerogenes,* and to a lesser extent between other species as well. In the methyl red test, a medium that contains a very small amount of carbohydrate fermentable by both organisms is inoculated with each of the organisms. Because typical strains of *E. coli* ferment carbohydrates, producing a variety of acids, a pH is obtained that is low enough to change methyl red to its acid color. Typical strains of aerogenes, utilizing this same quantity of carbohydrate, are not capable of producing so low a pH. Some of the end products formed by *A. aerogenes* are nonacidic, such as ethyl alcohol and acetyl methyl carbinol. Thus, using the same starting products, *E. coli* produces a series of predominately acid end products, while *Aerobacter aerogenes* produces only a partial assortment of acid end products. Therefore, the culture of *E. coli* produces a pH low enough to register an acid color (red) of methyl red, while *Aerobacter aerogenes* does not reach this pH and displays the alkaline color (yellow) of methyl red.

The second part, the Voges-Proskauer test, confirms this difference by showing that *E. coli* does not form acetyl methyl carbinol, whereas typical strains of *Aerobacter aerogenes* do. This is simply an ordinary chemical test to detect acetyl methyl carbinol in the medium. The combination of these two tests helps confirm the separate identification of these two organisms. These tests demonstrate very graphically the practical use to which a good knowledge of microbial respiration can be put. Once the end products of respiration of the two different organisms are known, a differential type of test may be used. When the pH is approximately 4.4 or lower, methyl red is red; when the pH is 6.0 or higher, methyl red is yellow.

EXPERIMENT 4-14

The MR-VP Procedure

MATERIALS

1. Cultures of Escherichia coli and Aerobacter aerogenes.
2. Tubes of MR-VP broth (4).
3. Methyl red indicator.
4. α-naphthol solution.
5. Potassium hydroxide solution.
6. Sterile pipets (2).

PROCEDURE: Inoculate two tubes of MR-VP broth with E. coli and two tubes with A. aerogenes and incubate at 37°C. for two days. After incubation, perform the methyl red test on one tube of each organism as follows: Add about five drops of methyl red indicator to each tube and record the resulting color displayed for E. coli and for Aerobacter aerogenes.

With the unused set of MR-VP broth tubes, perform the Voges-Proskauer test as follows: Add 0.5 ml. of α-naphthol solution to each of the tubes. Then add 0.5 ml. of the potassium hydroxide solution. Agitate and allow to stand in the laboratory for 1 to 2 hours. Look for the appearance of a pink to red color, which is a positive indication of the presence of acetyl methyl carbinol.

One enzyme that appears to have a strong correlation with the ability of an organism to utilize oxygen is called *catalase*. Catalase, which is responsible for the degradation of hydrogen peroxide into oxygen and water, is not found in anaerobic organisms nor in certain species that are microaerophilic. Apparently the lack of catalase in anaerobic organisms partially explains why oxygen is poisonous to anaerobes. When oxygen is available, these organisms apparently use it as a final hydrogen acceptor to form hydrogen peroxide, but anaerobes cannot degrade it any further, as other organisms can. Thus hydrogen peroxide accumulates and poisons anaerobes because of increased concentration. Catalase is very easily detected, as is demonstrated in the following experiment.

EXPERIMENT 4-15

Detection of Catalase in Bacteria

MATERIALS

1. Culture of Escherichia coli and Lactobacillus delbrueckii.*
2. Tryptose phosphate agar talls (2).
3. Hydrogen peroxide solution (3.0%).
4. Petri plates (2).

PROCEDURE: Do two streak plates for isolation on Petri plates of tryptose phosphate agar with both E. coli and L. delbrueckii. Incubate both plates at 37°C. for 48 hours. Flood the plates with the hydrogen peroxide solution and look for visible evidence of the presence of catalase in the organisms, which can be identified by a trail of bubbles arising from the colony submerged in the hydrogen peroxide. Record the results. What is the gas being evolved in this reaction? The reaction is summarized as follows:

$$2H_2O_2 \xrightarrow{\text{(catalase)}} 2H_2O + O_2 \uparrow$$
(hydrogen peroxide)

Lactobacillus delbrueckii is a rod-shaped organism normally found in fermenting green mashes and other vegetable matter.

CHAPTER 5

DESTRUCTION AND INHIBITION OF MICROORGANISMS

ENVIRONMENTAL AND PHYSICAL EFFECTS

There are many conditions in the environment of microorganisms that are lethal or potentially lethal for them, and one of the purposes of this section is to study some of these conditions and how they affect the viability of specific microorganisms. It is a general rule of thumb that microorganisms are usually more resistant to death by physical and environmental means than are the cells of higher forms of life. It is often very difficult for the beginning student to realize just how drastically the environment changes for a given microbial population from one moment to the next. Changes in temperature, osmotic pressure, and hydrogen ion concentration all occur many times with great rapidity. Cells of higher organisms are generally very intolerant of much change in any of these parameters. A tiny drop of moisture on a rock can support a thriving population of microorganisms until the temperature rises and the droplet suddenly completely evaporates. This illustration is typical of the adverse conditions constantly developing in the micro-environment.

Experiments in this chapter illustrate the general effects of physical and chemical forces of the environment upon the growth and survival of microorganisms. In the next experiment, for example, the effects of various hydrogen ion concentrations are considered relative to the suppression of microbial growth. In this experiment the broth into which the organisms are inoculated is nutritionally able to support growth of the organisms, provided the pH is tolerable. The only differences between the various inoculated tubes of media are in the pH of the broths. The student will draw his own conclusions from observations made on growth. The student should read the results from all tubes in the experiment at the same time, otherwise the comparison will not be valid.

EXPERIMENT 5-1

The Effect of Hydrogen Ion Concentration (pH) upon Microorganisms

MATERIALS

1. Cultures of Escherichia coli and Saccharomyces cerevisiae.
2. Tubes of sterile nutrient broth previously adjusted with 1.0 N HCl or 1 N NaOH to pH values of 3.0, 5.0, 7.0, 9.0, and 11.0 (1 each).
3. Tubes of sterile malt extract broth adjusted to these same pH values (1 each).

PROCEDURE: Inoculate the E. coli culture into each of the tubes of nutrient broth at the various pH values. Repeat this procedure with the S. cerevisiae in the malt extract broth at the various pH values. Incubate all tubes in your drawer and record your observations at one, two, and five days in terms of quantities of growth. The symbols for recording this exercise should be 4+ for maximum growth and 0 for no growth. Judge each intermediate tube as 3+, 2+, or 1+.

In several of the experiments to follow you will be asked to judge the amount of microbial growth in a liquid medium. Although the 4+ − 0 system is reasonably accurate, a more quantitative approach can be designed rather simply. If one mixes barium chloride with sulfuric acid, a precipitate of barium sulfate ensues. By adjusting the concentrations of the reactants it is possible to create varying amounts of precipitate. The McFarland Scale is reproduced below, in which each mixture of $BaCl_2$ and H_2SO_4 has been calibrated to approximate a certain number of bacteria per unit volume. In other words, the turbidity resulting when one of these tubes is shaken approximates the turbidity resulting from a certain number of bacteria. If you prepare a series of test tubes as described below and cap them well, you will have a permanent set of counting standards for approximating the population density of tubes of microbial growth. By mixing known and unknown tubes well, and then holding them side-by-side against a strong light, you should be able to estimate the amount of bacterial growth. Alternatively, if a spectrophotometer is available, one can utilize this instrument to give the same information even more accurately. In the latter case you must be sure to use an uninoculated broth tube to "blank" the microbial growth and a water blank for the barium sulfate tube.

McFARLAND SCALE

McFarland Scale	1% BaCl$_2$ (ml.)	1% H$_2$SO$_4$ (ml.)	Approx. No. Bacteria Represented x 10^6/ml.*
1	0.1	9.9	300
2	0.2	9.8	600
3	0.3	9.7	900
4	0.4	9.6	1200
5	0.5	9.5	1500
6	0.6	9.4	1800
7	0.7	9.3	2100
8	0.8	9.2	2400
9	0.9	9.1	2700
10	1.0	9.0	3000

*This range of $3 \times 10^8 - 3 \times 10^9$ bacteria per ml. is a useful one in that most of your growth experiments will fall within it, except for very poor growth.

One of the most frequently encountered conditions in the environment of microorganisms that might be considered detrimental to their welfare is a change in osmotic pressure. A microorganism in various natural environments is constantly faced with drastic alterations in the water content of the surrounding medium. When the concentration of dissolved materials surrounding an organism becomes very high in relation to the materials dissolved within the organism, the adverse condition of high *osmotic pressure,* or a *hypertonic* situation, results.

Osmotic pressure describes the tendency of water to move in the direction that equalizes the concentration of water in all parts of the system. In all cases where osmosis is involved, there is a semipermeable membrane separating one part of the system from another. In microbiology this usually refers to the cytoplasmic membrane that surrounds the cell, which is semipermeable. It therefore acts as a barrier between the interior solution of protoplasm and the external solution of the environment. When water concentration is the same on both sides of this membrane, a state of equilibrium is established and is said to be *isotonic.* In this condition, water moves equally in both directions, and there is no ill effect to the organism. If, however, the concentration of water outside the cell is less than that inside the cell, the tendency is for more water to flow out from inside the cell to equalize the water concentration on both sides of the membrane. This carried to completion would result in a shrinking of the solid contents of the cell. The system in which there is a shrinking of the cytoplasm as a result of loss of water to the surrounding environment is said to be *hypertonic.* On the other hand, if the concentration of water in the cell is less than the concentration of water outside the cell, there is a tendency for water to flow inward. In this situation, the cell swells as it takes in water. This system is said to be *hypotonic.*

Microorganisms also frequently find themselves in *hypotonic* situations, where the concentration of dissolved material outside the cell is much less than that inside the cell. Each group of microorganisms has its own special

DESTRUCTION AND INHIBITION OF MICROORGANISMS

Hypertonic
(Water diffuses
out of the cell)

Isotonic
(Water diffuses
equally in and
out of the cell)

Hypotonic
(Water diffuses
into the cell)

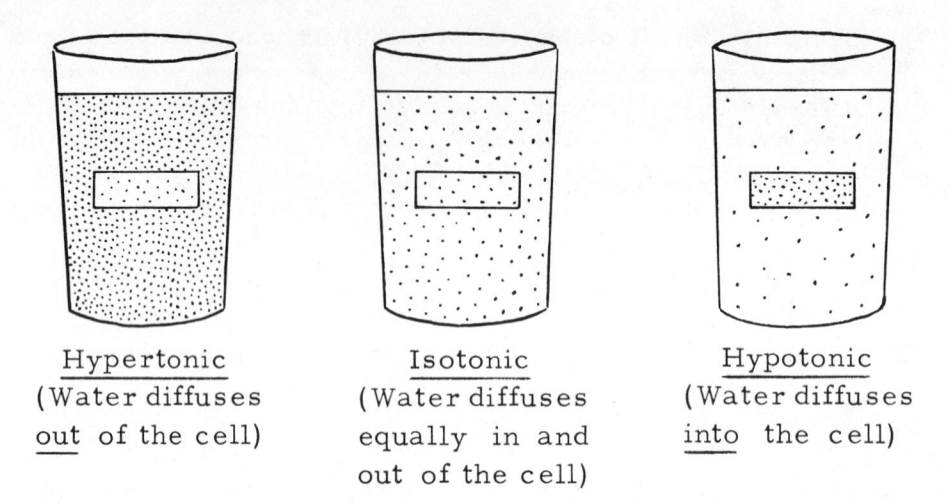

Figure 5-1 Diagrammatic representation of relative concentrations of dissolved particles (dots) inside and outside the cell in three different osmotic gradients.

way of dealing with these problems. Bacteria are not concerned with hypotonic situations particularly, because of their rigid cell wall, which prevents any bursting or *plasmoptysis*. Other microorganisms, such as the protozoans, have special contractile vacuoles that serve to eliminate the inward movement of water. In general, microorganisms and plant cells are not adversely affected by hypotonic solutions. In an isotonic environment, microorganisms are unaffected by osmotic pressure. However the picture becomes one of progressive damage to the cell as the system becomes more hypertonic, as is shown in the next experiment. Many practical applications are made of this fact, particularly in home or commercial canning, where such foods as jams and preserves are made exceedingly hypertonic and therefore need not be sterilized to prevent bacterial decomposition. Certain microorganisms, such as molds, are more tolerant of hypertonic solutions than bacteria and other microorganisms, hence this type of food can become overgrown with molds. For this reason, canned foods are ordinarily sealed so as to prevent contact with oxygen. Without oxygen molds do not grow.

EXPERIMENT 5-2

The Effect of Osmotic Pressure on Microorganisms

MATERIALS

1. A culture of Escherichia coli, and a spore suspension of Aspergillus niger (mold).
2. Tubes of nutrient broth adjusted to the following concentrations of glucose: normal (0.0%), 5.0%, 10%, 25% (1 each).
3. Tubes of malt extract broth adjusted to the following concentrations of glucose: normal (0.0%), 15%, 30%, 40% (1 each).

PROCEDURE: Inoculate one of each of the tubes of nutrient broth with E. coli culture and the tubes of malt extract broth with A. niger. Incubate both sets of tubes in your drawer and examine for the presence of growth during a period of one week. Record the results, using the system described in Experiment 5-1 (i.e., 4+, 3+, 2+, etc., or McFarland scale). Glucose is the key ingredient in both of the media employed, since its concentration is the only factor that is varied. Osmotic pressures are primarily based on the concentration of this substance in these media.

One of the physical agents known to affect growth of microorganisms (and all other forms of life) is radiation energy. Radiation can take many different forms, such as X-rays, beta rays, and gamma rays, as well as ordinary light. They differ from each other only in wavelength and energy content. It has been known for some time that light rays having a wave length of about 2600 to 2700 Angstroms (or 260 to 270 mμ)* are lethal to living cells in general, as well as to microorganisms. Light of this wave length is invisible, since we can see only those wave lengths between about 4000 and 7000 Ångstroms. The wave lengths in the 2600 to 2700 Ångstrom range are in the deep ultraviolet range. Light of this wave length is readily absorbed by certain cellular materials, particularly the nucleic acids, and apparently it causes irreversible damage if the time and intensity are sufficient. This is due to the fact that such radiation is sufficiently energetic to displace electrons from the affected molecules, thus ionizing them. The latter then become highly reactive chemically and enter into aberrant chemical combinations. Ionizing radiation is capable of causing permanent changes in the DNA (hereditary material) of a cell, such a change being known as a *mutation.*

This observation has resulted in attempts to control the growth of microorganisms with ultraviolet radiation. Ultraviolet lamps are often called sterilizing lamps. Unfortunately, ultraviolet radiation offers several important drawbacks as a sterilizing device. Outstanding among these is its inability to penetrate materials commonly found in the bacterial environment. To be killed by ultraviolet radiation, the bacterial cell must be directly exposed to the light and cannot be shielded by water, glass, or many other materials and objects. Any protection of this sort is complete, for ultraviolet light will not affect the organisms so shielded. Ultraviolet light is known to penetrate quartz, which can be made into glasslike sheets. This is used to advantage in such things as ultraviolet microscope lenses, which allow passage of these rays for optical purposes.

From a practical point of view, the killing of microorganisms with ultraviolet light is a very limited technique, since there are very few situations in which the organisms would be truly vulnerable to this type of

---

*Although this book will continue to use Å or mμ as units of radiation wavelengths, it is now becoming more accepted practice to employ nanometers (nm) for this type of measurement. A nanometer is equal to 10 Å and is therefore the same as a millimicron (10^{-9} meters).

radiation. Even in situations in which a room is to be sterilized (or at least the bacterial population reduced), most microorganisms would be protected from a single light source. Nevertheless, operating rooms are given considerable treatment with ultraviolet radiation prior to use so as to reduce the bacterial population in the air and the dust. The effectiveness of the light in killing is also related to both the distance between the organism and the light source, and to the time of exposure. Ordinarily the organism must be reasonably close to be affected at all.

In the following experiment, you will observe several aspects of the killing of microorganisms by ultraviolet radiation. If practical, it would be possible to set up similar experiments using other types of radiation, such as x-rays and various radiations from naturally radioactive elements. However, very strict precautionary measures are needed in using these radiations, so you will not work with them.

EXPERIMENT 5-3

The Effect of Ultraviolet Radiation on Microorganisms

MATERIALS

1. Culture of Escherichia coli (broth).
2. Nutrient agar for plating.
3. Ultraviolet lamp with a wave length of 2650 Ångstroms.
4. Petri dishes (3).
5. White paper.

PROCEDURE: Melt some nutrient agar and set up three Petri dishes with the agar in the usual way. After the agar has hardened, totally inoculate the surface of each dish with the culture of E. coli, using your bent glass rod. Next cut out a design in the center of a piece of white paper in any shape or form you desire, preferably one that is reasonably small and distinct, such as a star, a circle, or an initial. Make sure that the width of the paper exceeds that of the Petri dish. Remove the lid of the first inoculated Petri dish and replace the lid with this piece of paper. Place that paper so the design will be more or less centered over the Petri plate. Put the plate with the paper lid under the ultraviolet lamp at a distance of approximately 1 ft. and irradiate for 5 min. Caution: Be sure you do not look directly at the rays of this lamp. Also avoid looking at reflections of this light, for they too can damage the eye. Repeat the process on plate 2, except this time irradiate for 10 min. at the same distance using the same design. Expose plate 3 to the lamp under the same conditions for 10 min. except this time put the Petri dish lid on over the paper. After irradiation has been complete, replace the lids on all dishes and incubate for 24 hours. After incubation, observe your plates and record your conclusions both in tabular form and in a statement explaining the purpose of each of these three setups. Be sure that each Petri dish is properly labeled as to length of exposure and the conditions. As an optional experiment set up another plate and expose it with the lid off to direct sunlight for 2 hours or so to see how the plate is affected.

Another physical factor in the environment of the microorganism is temperature. All microorganisms can be classified according to the best (or optimum) temperature for growth. *Psychrophilic* microorganisms grow best between 0° and 20°C., *mesophilic* microorganisms grow best between 20° and 40°C., and *thermophilic* microorganisms grow best between 40° and 80°C. Therefore, when speaking of destructive temperatures, one must first be aware that these differences exist naturally among different microorganisms. In general, however, when we speak of the effect of temperature on microorganisms, we are referring to the lethal effect caused by the irreversible denaturation of intracellular proteins and other vital components of the cell. In this sense, then, temperature becomes one of those temperatures in the elevated ranges between 50° and 100°C. Naturally thermophilic bacteria (vegetative cells) are most tolerant to these temperatures. However, ruling out these organisms and spore-formers and considering only the others, one finds that in general the resistance to temperature is correlated primarily to hereditary factors in the cell. For example, some non-spore-forming organisms are more resistant to heat than other non-spore-forming types. In general, most bacteria are affected adversely by temperatures in excess of 50°C. although bacteria do differ considerably as to the length of time required to kill. In the experiment to follow, two types of bacteria are compared. One is an endospore-forming organism and the other is non-endospore-forming. The specific purpose of the experiment is to determine at a given temperature the time of exposure needed to kill all organisms in the culture.

EXPERIMENT 5-4

The Effect of Heat on Microbial Survival

MATERIALS

1. Slant cultures of <u>Bacillus subtilis</u> and <u>Escherichia coli</u>.*
2. Tubes of nutrient broth (2).
3. Centigrade thermometer (1) and water bath (1).
4. Clean test tube (1).
5. Petri plates (6).
6. Nutrient agar for plating.

---

*If time and facilities permit, it is often instructive to include *Staphylococcus aureus* as a third test organism in this experiment, so that it can be compared with another non-spore-forming bacterium in terms of temperature resistance.

PROCEDURE: Make a separate heavy suspension in sterile nutrient broth of B. subtilis and E. coli by inoculating a liberal loopful from a slant culture. Immerse both cultures in a water bath set at 50°C. by previous heating with a Bunsen burner. The water level should be slightly deeper than the broth in the tube. Maintain the water bath at 50°C. To determine if the contents of the tube are at the desired temperature, introduce into the water bath a third test tube that contains an amount of tap water approximately equal to the amount of broth culture in each of the test tubes containing organisms. Into the tube of tap water insert a centrigrade thermometer. When the temperature of the contents of the test tubes, as indicated by the centigrade thermometer, is equal to the temperature of the water bath, then you are ready to begin the experiment. While the temperature is rising to the desired level, prepare six plates with nutrient agar and divide or sector the bottoms of each plate into six equal parts. Label the parts as follows: "C" for control, "U" for uninoculated, 5 min., 10 min., 15 min., and 20 min.

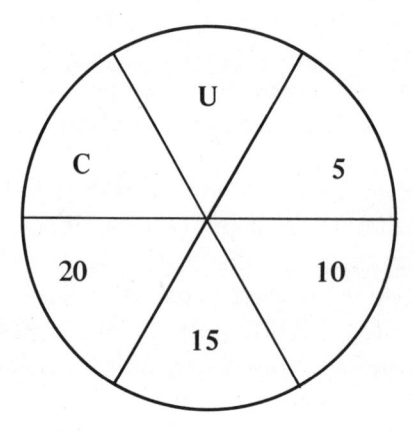

On the tops of three plates, label with the name "B. subtilis" and on the tops of the other three label "E. coli." At the beginning of the experiment, inoculate each sector marked "C" with the appropriate organisms as a viability check of each organism by making a single straight line with your inoculating loop through the center of the sector. Do not streak to the side of the sector, for a spreading growth may give confusing results. The sector marked "U" on all plates will remain uninoculated.

When all plates are ready and the temperature of the water bath has reached 50°C., using one plate labeled E. coli and one plate labeled B. subtilis, inoculate each sector marked "5 min." with the appropriate organism from the water bath after the organisms in the tubes have been exposed for 5 min. to this temperature. Be very careful that you do not touch the loop to any of the glass on the inside of the tube of bacteria, since bacteria on these surfaces might escape exposure to the desired temperature. Similarly after another 5 min., apply a single line of growth to the sector marked "10 min." Continue in the same manner until both plates are entirely inoculated.

Then gradually raise the temperature of the tubes in the water bath to 70°C. Repeat the entire procedure with the second set of plates at 70°C. Finally, repeat the procedure again after elevating the temperature to 85°C., so that you have three sets of plates, each set representing a different temperature. Be sure to mark the plates according to the temperature used. In this experiment look for the presence of <u>growth</u> or <u>no growth</u> along the line of streak after incubating all plates for 24 to 48 hours in the 37°C incubator. Record your observations as to whether the organisms grew under these conditions or not. What conclusions can you draw as to the relative resistance to heat on the part of these two organisms?

<u>NOTE</u>: Many students will find that the results of this experiment are not entirely orderly. You may find no growth at a lower temperature and growth at a higher temperature on the part of the same organism. If this should happen to you, attempt to determine the possible sources of error. To do this, be sure that you think over precisely what you are trying to prove.

PLANNED DESTRUCTION OF MICROORGANISMS

Since the advent of the germ theory of disease, one of the principal concerns of man has been the planned destruction of microorganisms in and around the human environment, when such measures are desirable. In most instances, this type of activity is aimed at the destruction of *pathogenic* (disease-causing) organisms. Many substances known to be toxic to living cells in general are used extensively for this purpose. The search for substances with high germicidal properties has been one of the most intensive of our times.

Many problems are associated with the practical use of such materials. Paramount among these is, of course, the fact that in many instances, these substances must be used in contact with living human cells. If the substance is just as toxic (or more so) to human cells as to microbial cells, no advantage is gained from its use. In general, one wishes to utilize a substance that is quite active in dilute solution, that is more toxic to bacteria than to human tissue, and that has a very rapid rate of action. In addition, the substance must not be inhibited by the presence of organic matter, such as is found in antisepsis (serum, blood, body fluids, etc.), and the substance must not be species specific in its action (i.e., it must be equally effective against all kinds of organisms). Other considerations of lesser importance indicate that it must be reasonable in cost, penetrating in action, and that it preferably should stain or mark the area on which it is applied. All of these characteristics are concerned with destruction of microorganisms in an external environment and have nothing to do with the killing of microorganisms within the body. For this purpose, antibiotics and chemotherapeutic agents are employed due to the lower tissue toxicity of these agents compared to antiseptics, but the aim is the same in either case — to rid the individual of potentially undesirable microorganisms.

In many subsequent experiments, a testing procedure is employed based on the diffusion of a substance in solution (agar is a solution even though solidified) from a point of *initial high concentration* into an area of *low* (or no) *concentration* of the same substance. When this diffusion occurs, the substance in question is most effective at the area of highest concentration; its effectiveness decreases as the concentration diminishes. The following experiment is designed to demonstrate *visually* the simple diffusion of several colored substances through agar, so you can obtain a better understanding of the subsequent experiments in which the diffusing material is usually invisible.

EXPERIMENT 5-5

Demonstration of Molecular Diffusion in Agar

MATERIALS

1. Nutrient agar for plating.
2. Aqueous solutions of methylene blue (0.5%), crystal violet (0.5%), and safranin (0.5%).
3. Filter paper discs (¼ in. diameter).
4. Petri dish (1).
5. Forceps (1).

PROCEDURE: Melt the nutrient agar and pour into a Petri dish in the usual way. Allow the agar to harden. Mark off the nutrient agar plate into three sectors. With forceps, take a filter paper disc, hold it over one of the solutions of dye, and dip just the edge of it into the dye. Observe that the dye rapidly climbs up the filter paper disc through capillary action. When the dye has reached about two-thirds of the way up the disc, remove the disc from the solution. The dye should continue to spread all the way across the disc. When the dye has completely impregnated the disc, deposit the disc in one of three sectors previously marked off on the nutrient agar plate. Place the disc fairly near the outside edge so that full diffusion of the dye toward the center of the plate can be observed. Press gently on the top of the disc with the forceps to insure firm contact of the disc with the agar. Repeat the same procedure with the other two dyes. Invert the plate and label each sector. Put the Petri dish in your drawer and observe after one day. Note that the dye in the agar is diluted (color is less intense) as it diffuses away from the disc.

In future experiments utilizing this technique, keep in mind that even though the diffusing substance may not be colored (and therefore invisible), the diffusion is essentially the same. One difference is that various substances exhibit different abilities for diffusing great distances. Make a drawing that illustrates what you have seen in this experiment. Keep in mind that the intensity of color in your plate is proportional to the concentration of the material. This is the principle upon which all such experiments are based.

Several of the dyes used for staining microorganisms are quite useful antiseptics in certain instances. These dyes are even more useful in adjusting conditions in a bacteriological medium, so as to favor or suppress growth of certain species. One dye used extensively in this respect is crystal violet (gentian violet). This dye at relatively high concentrations is inhibitory to both gram-negative and gram-positive organisms. However, when the concentration is reduced, the dye is selective in its action, in that it kills, or inhibits gram-positive organisms but not gram-negative organisms. Finally, a further reduction in concentration results in a level tolerable to both types of bacteria. When crystal violet is tested, utilizing the diffusion technique illustrated in the previous experiment, a very simple experiment is provided demonstrating the principles of selective action. Because of the high activity of crystal violet against gram-positive organisms, it is useful as an antiseptic in infections caused by gram-positive organisms. Crystal violet is particularly effective in throat infections where *Streptococci* and *Diplococci* are most likely to be involved. In addition, this substance is incorporated into bacteriological media to suppress the growth of gram-positive organisms, thereby making the medium selective for gram-negative organisms.

EXPERIMENT 5-6

The Selective Action of Crystal Violet

MATERIALS

1. Broth cultures of Staphylococcus aureus and Escherichia coli.
2. Sterile aqueous crystal violet solutions in the following concentrations: 1:10,000, 1:50,000, and 1:100,000.
3. Nutrient agar for plating.
4. Sterile Petri dishes (3).
5. Filter paper discs (sterile).

PROCEDURE: Be quite careful to maintain aseptic precautions during this experiment so materials remain sterile. Melt the nutrient agar and cool to approximately 45°C. Pour into sterile Petri plates and label one plate 1:10,000, the second plate 1:50,000, and the third plate 1:100,000. After the agar has hardened, divide each plate into two sectors with a line bisecting the back of the plate. With your inoculating loop, streak one half of each plate with E. coli liberally so that growth will be continuous. Streak the other half of each plate in the same fashion with Staph. aureus. This procedure necessitates back-streaking and cross-streaking to make sure that the bacteria are distributed over the entire surface of each sector. After each plate has been streaked, dip a sterile filter paper disc in the 1:10,000 solution of crystal violet as you did in the previous experiment. Place the dye-containing disc near the outer edge on the E. coli half of the plate, labeled with the correct concentration. Repeat the same procedure on the half of the plate inoculated with Staph. aureus. Place a filter paper disc soaked in the appropriate concentration of crystal violet in each sector of the two remaining plates. Invert all plates and incubate at 37°C. for 24 to 48 hours; observe the results for the two different organisms. Make drawings of your plates for both organisms showing the results of this experiment.

For many hundreds of years eating utensils have been made using silver, either as a coating or in the solid form. Long before the germ theory of disease was known, this practice was in effect. Today we know that there is an intelligent basis for this practice. The silver ion is one of the metallic ions that is exceedingly active in the inhibition of microbial growth, even when used in very dilute solutions. It would perhaps be surprising to the student to realize that solid silver, such as that found in eating utensils or in a coin, releases enough silver ions in an aqueous solution to exert such activity. The same is also true of other heavy metal ions, such as copper, although silver is more active. Since our coins no longer contain appreciable quantities of silver, copper will have to be the metal of choice to demonstrate this effect in the following experiment.

EXPERIMENT 5-7

The Oligodynamic Action of Heavy Metals

MATERIALS

1. Cultures of Escherichia coli and Staphylococcus aureus (broth).
2. Nutrient agar for plating.
3. Two well-cleaned pennies.
4. Petri dishes (2).
5. 1 N hydrochloric acid — 95% alcohol solution.
6. Distilled water.
7. Nutrient agar shorts (2).
8. Forceps (1).

PROCEDURE: Prepare two Petri plates with nutrient agar in the usual way. Melt two nutrient agar shorts and cool to 45°C. Inoculate one of the shorts with E. coli and the other with Staph. aureus using about two loopfuls for each inoculum. Rotate these tubes of inoculated agar between the palms of your hands to distribute the organisms evenly throughout the agar. Maintain the two agar shorts at 45°C. while you proceed with the rest of the experiment. Thoroughly clean a penny by first washing with soap and water; then dip it into the acid-alcohol cleaning solution with a pair of forceps, and rinse with distilled water. Be sure to handle the penny with forceps. Flame the forceps before use.* After the coin has been rinsed with distilled water, place it firmly in the center of the agar plate so that it adheres to the agar surface. Then pour the melted, cooled, and inoculated agar short of E. coli over the penny so that a second layer of agar is formed above the first. Close the plate, label with organism name, and incubate inverted at 37°C. for 24 to 48 hours. Repeat the same procedure with the other Petri dish, using the Staph. aureus, and label. Observe results and make drawings illustrating the activity of the copper ion as illustrated by this experiment.

An experiment of more than usual interest to the microbiology student can help him evaluate the commercial products about which he hears and reads. It is very difficult to pinpoint any single test as being completely valid in testing antiseptics for their overall effectiveness. Each of the so-called desirable qualities sought in an antiseptic should be tested for, and no one test is capable of covering all of these points. For practical reasons it is very difficult to have the student perform tests such as the Phenol Coefficient Test and the Salle Toxicity Index in the laboratory. However, there are a few simple tests that can give a partial answer. One of the easiest tests to perform is the filter paper disc method used in the previous two experiments, applying it to products that are commercially available. In the experiment to follow, you may use any four products that claim germicidal properties, ranging from toothpaste to antiseptics.

---

*Note: Do not flame the penny. It melts!

EXPERIMENT 5-8

The Filter Paper Disc Method of Evaluating Proprietary (Commercial) Antiseptics

MATERIALS

1. Broth cultures of Escherichia coli and Staphylococcus aureus.
2. Talls of sterile nutrient agar (2).
3. Four proprietary antiseptics of the student's own choosing. Several of these will also be provided in the laboratory.
4. Sterile filter paper discs.
5. Sterile peptone solution (10%) or blood plasma (sterile).
6. Petri dishes (2).

PROCEDURE: Prepare one agar plate each of E. coli and Staph. aureus in which these organisms are well distributed throughout the agar by inoculation of the melted, then cooled talls (2 loopfuls in each tall). Label each plate appropriately. After the plates have hardened, set up a filter paper disc method for evaluation of the four different antiseptics in each plate. Be sure you do not let the discs become oversoaked with the solutions you are testing, or the experiment may be ruined. Be careful to maintain strict aseptic conditions throughout this experiment. You might like to do one set of four antiseptics while your partner does another set of four. There is no limit to the number you may test, but be sure you test at least four. Be sure to expose both organisms to each antiseptic. Why? If you do not bring any of your own, you may perhaps borrow from your neighbors. In addition several varieties are provided in the laboratory for your use. Incubate the plates in the 37°C. incubator for 24 to 48 hours and observe your results. In your conclusions be sure to specify what this test shows and what it does not show in terms of evaluating the antiseptics you used.

There is a variation of this technique that can easily be done in this laboratory. Set up the experiment exactly as you did before, except this time into your melted and cooled agar pipet 2 ml. of a sterile peptone solution or blood plasma provided for you. Compare the results obtained with this second method, which should illustrate the relative activity of the substances under test in the presence of organic matter. Although blood plasma is a more realistic additive, it is not always available, but in most cases a concentrated peptone solution gives similar results. Be sure to include in your conclusions observations based on the modification of the test you have just performed.

Many times the physician, when treating infection, wishes to know not only which organism is involved, but also what antibiotics he may use to manage the infection. Many antibiotics lose their effectiveness against certain kinds of bacteria (due to bacterial mutation), and therefore it becomes very necessary to test each specific infection, using organisms taken from that infection, against a series of potentially effective antibiotics. Even if the identity of the organism is known, antibiotic susceptibility tests must be performed since so many mutant strains are in existence. In the following experiment, the principles utilized are exactly the same as those in the experiments you performed on the evaluation of disinfectants and antiseptics. The only difference is the testing of antibiotics instead of antiseptics. Generally speaking, antibiotics are suitable for use in the bloodstream, whereas antiseptics are restricted to surface use only (skin & mucous membranes). Disinfectants are used on inanimate objects.

EXPERIMENT 5-9

The Evaluation of Antibiotics by the Filter Paper Disc Method

MATERIALS

1. Broth cultures of Staphylococcus aureus, Escherichia coli, and Micrococcus luteus.
2. Sterile paper filter discs impregnated with penicillin, streptomycin or dihydrostreptomycin, tetracycline, and chloromycetin (all commercially available).
3. Forceps (1).
4. Talls of nutrient agar (3).
5. Petri dishes (3).

PROCEDURE: Set up three Petri plates in the same way as the previous experiment, in which two loopfuls of the appropriate organisms are dispersed throughout melted and cooled nutrient agar, which is subsequently poured into three sterile Petri plates. Label each plate according to the organism inoculated and section the back of each plate into four parts, one for each of the four antibiotics to be used. Using aseptic precautions, carefully place one of each of the four filter paper discs soaked in antibiotic onto the appropriate sector in each of the three Petri plates. Incubate from 24 to 48 hours, make sketches of your results, and give your conclusions regarding the usefulness of the various antibiotics for the different organisms.

One of the most frequently employed methods for evaluation of antiseptics involves an experimental method that determines the *rate of activity* of the antiseptic in question, testing the antiseptic at the strength it is customarily used. In this experiment the organisms subcultured from a tube containing a mixture of the bacteria plus the antiseptic in question do not grow if the antiseptic has killed them all; but if even a few organisms remain alive, the subculture should show growth, indicating failure of the antiseptic to kill *all* organisms. Such an experiment involves careful attention to detail, particularly timing and basic laboratory technique. In subculturing, the student should be very careful not to touch the inside of the tube containing antiseptic and bacteria with the inoculating loop. Furthermore, the time measurements should be reasonably precise so that they will represent a sure measure of the rate of action of the substance under test. In order to test several different antiseptics, a lengthy experiment is needed. Preplanning of all details and movements is essential. In the experiment to follow, you will test three different kinds of antiseptics to determine their rate of action, using a nutrient broth solution for incubation and subculture. You will then repeat the experiment, substituting a peptone-rich broth for nutrient broth. This should give you some idea about the effect of organic matter on the *K value,* or rate of action, of the substance under test. The volumetric ratio employed in this experiment is one part bacterial suspension to four parts of the chemical or antiseptic in question, resulting in a 1:5 dilution of the bacteria. In other words, the 1 ml. of culture plus 4 ml. of antiseptic result in an environment in which there is a considerable excess of the chemical. This is probably somewhat in excess of the ratio that would be expected under real conditions of antisepsis. In this experiment, a number of different substances are made available to you for testing. In addition, you will be asked to test a single proprietary (commercial) antiseptic of your own choosing. Two or more students should work together on this experiment.

EXPERIMENT 5-10

Determination of the Rate of Action of Antiseptics With and Without Organic Matter Present

MATERIALS

1. A nutrient broth culture and a peptone-rich nutrient broth culture of Staphylococcus aureus.
2. Tubes of nutrient broth (12).
3. Antiseptics provided in the laboratory in strengths commonly used.
4. A proprietary (commercial) antiseptic of the student's own choice.
5. Tubes of peptone-rich nutrient broth (12).
6. Sterile empty test tubes (6).
7. 10 ml. pipet (6).

PROCEDURE: Set up four tubes of nutrient broth labeled: "1 min.," "2 min.," "4 min.," and "8 min." Pipet 4.0 ml. of one antiseptic into a sterile empty test tube, using aseptic precautions. Then carefully pipet 1.0 ml. of the nutrient broth culture of Staph. aureus into the same tube and mix well after closing. Note the time immediately upon mixing. At each time interval (just mentioned), transfer two loopfuls of this mixture into a subculture tube of nutrient broth. In other words, after 1 min., subculture two loopfuls of the antiseptic plus bacteria suspension in the first tube; 2 min. after mixing, subculture in the tube marked "2 min.;" and so on until all four subculture tubes have been inoculated. Do all procedures at room temperature (approximately 20°C.)

Repeat the procedure with the second antiseptic provided. Finally, repeat the procedure a third time with a proprietary antiseptic of the student's choice. Repeat all tests on the three antiseptics, except this time use a culture of Staph. aureus that has been grown in peptone-rich nutrient broth and subculture into peptone-rich nutrient broth instead of regular nutrient broth. This will give you some idea of the effect of organic matter on the rate of killing of the three tested antiseptics. Incubate all tubes for 24 hours, and record your observations as to growth or no growth in each of the tubes. Carefully tabulate your results and write up your conclusions as to what the results showed.

One of the most difficult ideas to put across to the beginning student in microbiology is the omnipresence of microorganisms in his own environment, even after taking so-called effective measures for their control. One of the most important methods of control is skin antisepsis and cleanliness. In the following experiment, the student will observe the effect of soap-and-water treatment on the bacterial flora of his skin, as well as the effect of antiseptics in ridding the skin of microorganisms.

EXPERIMENT 5-11

Some Observations on Skin Cleanliness and Skin Antisepsis

MATERIALS

1. Nutrient agar for plating.
2. Sterile swabs (3).
3. Scalpel or razor blade (sterile).
4. Antiseptic solution.
5. Sterile Petri plates (3).
6. Sterile tube of either water or broth (1).

PROCEDURE: Pour three plates of nutrient agar and allow them to harden. Divide each plate into two halves. Label these halves consecutively, one through six. In sector 1, touch the agar with several of your fingers, then close the plate. Wash your hands with soap and water and while they are still wet (shake off the excess water), touch sector 2. Wash your hands a second time and touch sector 3, and then again for a third time and touch sector 4. On another area of skin apply the antiseptic provided using a sterile swab. Allow the antiseptic to remain for about 2 min. Then take another sterile swab, previously dipped in either sterile water or sterile broth, and wash over the area treated with the antiseptic, making sure that this area is moist. Transfer a little of the moisture from the swabbed area onto zone 5 using a clean swab. Allow the antiseptic to act on your skin for another 2 min., then scrape your skin gently with a previously sterilized scalpel or razor blade to dislodge some of the dead surface skin. Transfer some of the flakes of the skin to sector 6 and streak over the surface with the inoculating loop. The scalpel or razor blade may be sterilized for use in this exercise by soaking it in alcohol and flaming it gently. Allow it to cool before using. Incubate the plates from 24 to 48 hours at 37°C. and tabulate your results in terms of relative amounts of growth. Be sure to include your conclusions about the results of these observations. The results you obtain on the subcultures from your washing procedure may surprise you at first, but remember that most skin surfaces (and particularly the hands) are covered with a film of oil or grease which is not soluble in water. Soap breaks down and removes this barrier. In hospital practice, it is customary for the surgeon to scrub his hands for 10 min., using a brush followed by the application of an antiseptic to the skin prior to putting on sterilized gloves. Does this exercise help you understand why such a procedure is necessary?

CHAPTER 6

IDENTIFICATION OF UNKNOWN BACTERIA

By now it should be obvious that there are many different ways of characterizing microorganisms. In general, you have covered the cytological and biochemical aspects of characterization. Although your work thus far does not represent a complete study by any means, it is nevertheless sufficient to allow you to test your skill on an independent basis. Probably no other way of doing this is as satisfactory as the identification of an assigned unknown bacterium. The problem is laid out in an extremely simple fashion, but solving it requires the student to bring together all the skills he has acquired thus far. You will be given an organism identified by a code number only, and you will be asked to determine its genus and species. Such a procedure should logically require some knowledge of microbial classification.

The science of classification, or *taxonomy*, deals with the orderly arrangement of living things into intrarelated groups. For example, man is classified as a member of the Animal Kingdom, and with genus and species names of *Homo* and *sapiens* respectively. All living things classified and named under this binomial system of nomenclature are given a genus and a species name. *Families* consist of groups of similar *genera, orders* consist of similar families, *classes* consist of similar orders, and *divisions* contain similar classes. Each of the major *kingdoms* is divided into divisions. Although there is considerable evidence to indicate that microorganisms in general could be classified in a kingdom of their own (the protists), separated from both the plant and animal world, the traditional classification system continues to put most microorganisms in the first division of the Plant Kingdom, called *Protophyta* (primordial plants). Three major classes are included in the Protophyta — the *Schizophyceae* (blue-green algae), the *Schizomycetes* (fission fungi), and the *Microtatobiotes* (smallest life). Further supplementary details of the classification of microorganisms are given in your textbook or in "Bergey's Manual of Determinative Bacteriology."

In general, most of the organisms we will be concerned with fall in the fourth order, Eubacteriales (the *true bacteria*). Thirteen families in this order have been extensively considered in this book. It would be well to go through a general classification of microorganisms and locate the taxonomic niches for the various organisms you have already become familiar with during the semester. The more you do this, the easier the classification scheme will become for you. In the experiment concerned with identifying an unknown organism, the names given in the seventh edition of "Bergey's" will be used exclusively.

At this point, it would be well to mention the basis upon which biological classification is made. The underlying and unifying theme behind biological science is the concept of *organic evolution*. This concept relates all living things to each other, given a suitable passage of time. In essence, the evolutionary doctrine states simply that all existing life is derived from pre-existing forms. From studies of various types of life, it has become evident that transistions from one type of organism to another have been gradual, and therefore in classification it has become a useful tool to put living things into groups that appear to have an evolutionary relationship. Those organisms most closely related are those that are the closest on the evolutionary tree. Expanding this idea somewhat, the larger taxonomic groups are the larger branches of the evolutionary tree. The entire evolutionary tree of life is included in the classification of all living things.

Various clues lend themselves to classification. For example, the *appearance* of an organism is certainly a strong indication of relationship to a similar one. Thus the so-called *morphology* of an organism can be used as a tool for classification. In addition, the *method of reproduction* and other such observable facts can be used to strengthen the morphological approach to classification. In summary, it can be seen that we attempt to classify organisms on the basis of observable evolutionary relationships to each other.

Unfortunately, this method is very difficult to follow when classifying microorganisms, largely because they are so simple in structure that there is a minimum of morphological differentiation possible. In addition, the modes of reproduction are either identical or very similar. Because of this, microbiologists long ago were forced to utilize a "key" type of classification, in which specific observable characteristics on the biochemical level, as well as available morphological difference, were utilized to differentiate one kind of organism from another. This type of classification was developed largely under the direction of the American microbiologist David Hendricks Bergey. It was felt at the time that this system was an inferior substitute for the natural or evolutionary method and that it was, in general, an artificial approach to classification. However, in light of more recent knowledge of the relationships existing between the hereditary apparatus and the biochemical substances it produces, this approach to classification is probably as accurate and acceptable as the traditional approach. We now know that all organisms possess a nucleic acid information molecule (usually DNA) which encodes all of the units of heredity or "genes" of an organism. The genes direct the manufacture of an effector-type nucleic acid (RNA), and the latter directs the manufacture of all proteins within the cell. The proteins, primarily the enzymes, then direct all further activities of the cell, and ultimately the organism. Thus, relatedness, reduced to chemical terms, merely denotes the relative number of genes and/or proteins that any two organisms have in common, i.e., the more of the same genes or proteins they contain, the more related they are. As you have noted by now, many of the "key" tests you have performed are simply identification of specific enzymes (proteins) of microorganisms. This level of identification is closer to the gene level than gross appearance (morphology), and is less subject to error. It is easier to be mistaken when trying to establish the relationship between two organisms on the basis of their appearance than on the basis of the substances their cells manufacture — that is, chemical similarity is more fundamental than morphological similarity. Suffice it to say that at this time it appears biochemical analysis of any organism may very likely become the most acceptable way of determining its evolutionary relationships with other organisms.

Returning to a consideration of microbial classification, we can summarize by saying that in general it is a classification based on various types of biochemical analysis. For instance, the gram stain is widely employed to differentiate organisms. The gram stain alone does not identify an organism but assists in doing so. It was once thought that use of the gram stain was an artificial approach to classification, but it is now recognized that a gram-positive organism differs from a gram-negative organism because it contains biochemical materials and structures within its cytoplasm that are not found in the gram-negative organism. Furthermore, these materials are manufactured by the synthetic processes of the cell in question under direct control of the hereditary mechanisms of that cell. In other words, the genes of an organism control the manufacture of all substances within it, including those that make it gram-positive. Therefore, the gram stain is a valid aid to effect the differentiation of various organisms.

In addition to various stains, all of which indicate some hereditary characteristic of the organism, we utilize what might be considered an *enzymatic analysis* to assist in classification. Each enzyme, being a specific protein, is manufactured under genetic control. If an organism can hydrolyze starch, for example, we know that it produces the starch-hydrolyzing enzyme, and this substance can only be produced if the organism possesses the proper gene(s) to control its manufacture. Reflecting back on the work you have done in the laboratory, it should be easy for you to pinpoint a relationship between characteristics exhibited by the organisms you worked with and their heredity. Even such things as relative susceptibility to a given disinfectant or antibiotic are due to the heredity of the organism. The ability to utilize various sources of nitrogen and carbon in the manufacture of cellular materials likewise is a function of the presence or absence of specific enzymes and is again easily related to the heredity of the organism.

With these thoughts in mind, it now becomes a simple matter to consider the classification that has been devised for microorganisms. The book in which most of the data useful in the classification of various microorganisms has been compiled is "Bergey's Manual of Determinative Bacteriology." In this manual, one will find all of the known classified species of bacteria and other related forms along with a compilation of the specific characteristics of each organism. In order to make it easier for the student to find the organism he is attempting to identify, a system of keys has been devised in this manual. A key is a "follow-your-nose" type of thing in which you determine certain characteristics of the organism in question, and you then follow through the keys until you can tentatively identify the organism. Your instructor will explain this system further.

A *modified* and *abridged* version of some of the keys from Bergey's has been compiled and included in this laboratory textbook. This key can lead you to a tentative identification of all organisms you might be assigned. It has been phrased in the same fashion found in Bergey's manual, but it is reworded and shortened so only a few families of bacteria are included. After you have found your way to a specific genus and species by using this key, there are several pages provided listing all the possible organisms along with most of their specific characteristics, all itemized in chart form. Refer to this chart to complete your identification. In addition, a short descriptive key to the various families of bacteria is included for your convenience. After you have completed the identification up to this point, you may then compare your results with information in Bergey's manual. Although the facts in your charts were taken from Bergey's manual, your charts are not as complete as the information in Bergey's. Therefore you might wish to reassure yourself by a final check in Bergey's. However, under no circumstances should you use Bergey's for anything but a final check. Copies of Bergey's manual will be placed on reserve in the library. Be sure you refer only to editions dated 1948 or later.

A frequent problem that arises in determining the identity of an unknown bacterium is the inability of the student to determine whether he is observing a rod or a coccus. It should be re-emphasized that many rods assume a coccus-like form in a stained preparation. However, there will always be in the field of view a *few good distinct rods* that positively identify the culture as a bacillus rather than a coccus. The student should refer back to the section on cytology for additional discussion on the problems associated with coccobacilli. Students frequently ask the instructor during the course of this experiment to verify whether he has a gram-negative or a gram-positive organism, a rod or a coccus, and many other such interpretations. It should be made quite clear that this exercise *must* be carried on *entirely independently* by the student and that no hints or assistance will be given by the instructor. After you have received your assigned culture, simply go through the process of identification as best you can, keeping a record on the chart provided. Turn in the chart with all data you have accumulated, and the name of the organism written on it. This will constitute your laboratory report for this experiment. If you are wrong, and if time permits, you will be allowed to try again.

Before you actually begin your identification procedure, perhaps one of the difficulties that might be encountered should be mentioned. Without doubt, the most important problem in this type of experiment is *contamination*. It does you no good to identify a contaminant. To minimize the possibility of contamination, the student should take all necessary precautions not only to avoid contamination, but also to make it possible for him to recover the assigned organism again if contamination does occur. The procedure used in the following experiment helps to minimize this difficulty.

EXPERIMENT 6-1

Identification of an Unknown Bacterium

MATERIALS

1. Assigned culture of unknown bacterium (coded by number).
2. Media necessary (see Appendix).
3. Sterile Petri plates.
4. Sterile pipets, 1.0 ml.
5. Reagents for performing tests indicated (see Appendix).
6. Tubes of sterile distilled water, 1.0 ml. (1).

PROCEDURE: As soon as you have received your assigned culture, streak a Petri plate for isolation. To insure isolation, subdilute your inoculum in a small sterile distilled water blank before streaking the Petri dish.

As soon as the Petri dish has been streaked, do several gram stains on the assigned culture so you can immediately determine the predominating organism, even if a few contaminants happen to be present. In addition, prepare a wet mount and determine whether the organism is motile or not.

After your Petri plate has incubated a suitable length of time, pick a well-isolated colony with your inoculating needle, and streak the surface of two nutrient agar slants from the same colony. Label one tube as "reserve stock culture" and the other as "working stock culture." Incubate these until a reasonable amount of growth is obtained, and then keep them in your drawer. Do all of your subsequent testing and experiments with the working stock culture, and leave the reserve stock culture completely alone. The reserve stock culture will provide a new start on the organism if your working stock culture should become contaminated through repeated use. You will not be given a new start on your assigned culture, so treat the one you have with great care.

At least once every seven to nine days, a new set of stock cultures should be planted. Each time make two stock cultures as you did in the beginning, after streaking for re-isolation from your old reserve stock culture.

Each growth experiment should be checked with a gram stain on the growth from the tube itself (or from the Petri plate) to make sure your organism is growing and not a contaminant. In other words, if you inoculate a tube of lactose broth and get the formation of acid in such a tube (or any other result), make a gram stain on some organisms from that tube to make sure they still correspond to the organisms you started with. Each and every test should be checked in this fashion if your are to be absolutely sure. Even this type of precaution is not absolutely certain, since you can introduce a contaminant that resembles the one you were assigned. However, if these procedures are followed faithfully, the problems associated with contamination will virtually be eliminated, providing good laboratory technique is used.

Experience has shown that this exercise is one of the most enjoyable and profitable from the student's point of view. Consider it a challenge to your care and precision in the laboratory. The laboratory will be made available for your use at least a few hours every day of the week. It is highly advisable for you to conduct this exercise on a daily basis rather than just during assigned laboratory hours.

ABRIDGED KEY FOR IDENTIFICATION OF SOME COMMON BACTERIA*

Family Achromobacteriaceae

Small to medium straight rods; gram-negative; motile by means of peritrichous flagella, or nonmotile. May produce acid but no gas from glucose, but rarely ferments other carbohydrates. Considered to be poor fermenters.

KEY TO THE GENERA OF THE FAMILY ACHROMOBACTERIACEAE

I. Not active in the fermentation of sugars, especially lactose.
   A. Do not produce pigments on ordinary agar media.
      1. Litmus milk alkaline and no acid from carbohydrates.
Genus Alcaligenes.

KEY TO THE SPECIES OF GENUS ALCALIGENES

I. Gelatin not hydrolyzed (liquefied).
   A. Motile, aerobic.   *Alcaligenes faecalis.*
   B. Nonmotile.
      1. Produces ropiness in milk, good growth on nutrient agar, aerobic.
Alcaligenes viscolactis.
      2. Poor growth on nutrient agar, facultative.
Alcaligenes metalcaligenes.

---

*Information derived from the seventh edition of "Bergey's Manual of Determinative Bacteriology."

II. Gelatin liquefied.
      A. Motile.
         1. Milk peptonized (hydrolyzed). *Alcaligenes bookeri.*
         2. Milk not peptonized. *Alcaligenes recti.*
      B. Nonmotile. *Alcaligenes marshallii.*

Family Bacillaceae

Straight gram-positive (or gram-variable) spore-forming rods. Motile by peritrichous flagella, or nonmotile. Generally proteolytic; and are good fermenters, producing acid and occasionally gas from carbohydrates.

KEY TO THE SPECIES OF THE GENUS BACILLUS

   I. Aerobic or facultative, occur in chains, catalase positive.
      A. Spores do not swell the vegetative cell appreciably. Diameter of rods is 0.9μ or more.
         1. Acid from mannitol, Voges-Proskauer positive, slow liquefaction of gelatin, nitrate negative. *Bacillus megaterium.*
         2. No acid from mannitol, rapid liquefaction of gelatin, nitrate positive. *Bacillus cereus.*
      B. Diameter of rods is less than 0.9μ.
         1. Nitrites produced from nitrates, starch hydrolyzed, acid from mannitol. *Bacillus subtilis.*
      C. Spores swell the vegetative rod, gram-variable.
         1. Gas from carbohydrate fermentation, Voges-Proskauer positive, acid and gas from mannitol. *Bacillus polmyxa.*

Family Corynebacteriaceae

Straight gram-positive rods, often showing irregular staining segments and clublike swellings. Nonmotile, usually aerobic. Do not grow luxuriantly on common laboratory media. Often contain metachromatic granules.

KEY TO THE SPECIES OF THE GENUS CORYNEBACTERIUM

   I. Nonmotile.
      A. Acid from glucose. Nitrites not produced.
         1. Grows on ordinary agar. *Corynebacterium xerosis.*
      B. No acid from carbohydrates.
         1. Nitrites produced from nitrates.
            Corynebacterium pseudodiphtheriticum.

IDENTIFICATION OF UNKNOWN BACTERIA

Family Enterobacteriaceae

Straight rods, motile by peritrichous flagella, or nonmotile. Irregular arrangement. Gram-negative. Grow well on artificial media. All species attack glucose, producing acid, or acid and gas. Most species produce nitrites from nitrates. Most species are quite active fermenters of carbohydrates, frequently with the production of gas. Many species are normal inhabitants of the intestine of man and animals, while some are pathogenic to plants.

KEY TO THE TRIBES OF THE FAMILY ENTEROBACTERIACEAE

I. Lactose fermented, usually within 48 hours.
  A. Prodigiosin (red pigment) not produced.
    1. Not plant pathogens.   *Escherichieae.*
    2. Plant pathogens.   *Erwinieae.*
  B. Prodigiosin produced (red or orange pigment).   *Serratieae.*
II. Lactose not customarily fermented.
  A. Urea decomposed within 48 hours.   *Proteae.*
  B. Urea not decomposed within 48 hours.   *Salmonelleae.*

KEY TO THE GENERA OF THE TRIBE ESCHERICHIEAE

I. Lactose is fermented within 24 to 48 hours.
  A. Acetyl methyl carbinol not produced; methyl red positive; salts of citric acid may or may not be used as sole sources of carbon.
    Escherichia.
  B. Acetyl methyl carbinol produced; methyl red negative; salts of citric acid used as sole sources of carbon.
    1. Acid and gas from maltose, may or may not be encapsulated.
    Aerobacter.
    2. Acid only from maltose, heavily encapsulated.   *Klebsiella.*

KEY TO THE SPECIES OF THE GENUS ESCHERICHIA

I. Citric acid and salts of citric acid are not utilized as sole sources of carbon. Hydrogen sulfide not produced.
  A. Usually not pigmented, although a yellow pigment is sometimes produced.   *Escherichia coli.*
  B. Golden brown to red pigment produced.   *Escherichia aurescens.*
II. Citric acid and salts of citric acid are utilized as sole sources of carbon.
  A. Hydrogen sulfide produced.   *Escherichia freundii.*

KEY TO THE SPECIES OF THE GENUS AEROBACTER

I. Glycerol fermented with the production of acid and gas. Gelatin not liquefied (rarely liquefied). *Aerobacter aerogenes.*

II. Glycerol fermented with the production of no visible gas. Gelatin liquefied. *Aerobacter cloacae.*

KEY TO THE SPECIES OF THE GENUS KLEBSIELLA

I. Acetyl methyl carbinol produced. Frequently associated with acute inflammations of the respiratory tract. *Klebsiella pneumoniae.*

KEY TO THE SPECIES OF THE GENUS ERWINIA

I. Pathogens that cause dry necroses, gall or wilts in plants, but not a soft rot. Liquefy gelatin. Motile. *Erwinia amylovora.*

KEY TO THE SPECIES OF THE GENUS SERRATIA

I. Pigment not especially water soluble, readily soluble in alcohol.
   A. No visible gas from glucose.
      1. Inconspicuous pellicle, if any, on plain gelatin.
         Serratia marcescens.
      2. Brilliant orange-red pellicle on plain gelatin. *Serratia indica.*

KEY TO THE SPECIES OF THE GENUS PROTEUS

I. Urea hydrolyzed.
   A. No acid or gas from mannitol.
      1. Acid and gas from maltose. *Proteus vulgaris.*
      2. No acid or gas from maltose.
         a. Indole not produced. *Proteus mirabilis.*
         b. Indole is produced. *Proteus morganii.*

Family Lactobacillaceae

Gram-positive. Organisms extremely active fermenters. Some species are rods and others are cocci. All organisms tend to show characteristic grouping in pairs, or in short or long chains. Catalase negative. Most species grow with difficulty on the surface of ordinary laboratory media and are considered microaerophilic to anaerobic. Some species are pathogenic. Usually non-motile, but may be motile by peritrichous flagella.

KEY TO THE SPECIES OF THE GENUS LACTOBACILLUS

I. Rods occurring singly, in pairs, and in chains. Individual cells may be very long or even filamentous. Microaerophilic to anaerobic.
    A. Optimum temperature, between 37° and 60°C. or higher.
        1. Produce acid from lactose. Mannitol not fermented.
            Lactobacillus lactis.
        2. Does not produce acid from lactose. Mannitol not fermented.
            Lactobacillus delbrueckii.
    B. Optimum temperature, between 28° and 32°C
        1. Often prefers lactose to sucrose and maltose. Mannitol fermented.   *Lactobacillus casei.*

Family Micrococcaceae

Gram-positive or gram-variable, nonmotile cocci. Cell division produces characteristic groupings, which may be either irregular clusters, pairs, tetrads, or packets of eight. Many species form pigments and most of them grow readily on ordinary culture media. No gas is produced from carbohydrates, but they are good fermenters.

KEY TO THE GENERA OF THE FAMILY MICROCOCCACEAE

I. Aerobic to facultatively anaerobic species. Also includes some obligate anaerobes that occur in packets (Sarcina).
    A. Cells generally found in irregular masses; occasionally cells are single or in pairs.
        1. Action on glucose, if any, is oxidative. Aerobic. Some species gram-variable or gram-negative.   *Micrococcus.*
        2. Glucose fermented anaerobically with the production of acid. Facultatively anaerobic.   *Staphylococcus.*
    B. Cells normally occur in tetrads (fours) or packets of eight cells.
        1. Parasitic species occurring in tetrads. White to pale yellow chromogenesis. Nonmotile.   *Gaffkya.*
        2. Cells occur in packets, usually eight per packet. White, yellow, orange, and red chromogenesis. Usually nonmotile.   *Sarcina.*

KEY TO THE SPECIES OF THE GENUS MICROCOCCUS

I. May or may not reduce nitrates to nitrites, no gas produced in carbohydrates.
    A. No pink or red pigment produced on agar media in young cultures.
        1. Nitrites not produced from nitrates.
            a. Utilize $NH_4H_2PO_4$ as a sole source of nitrogen. Yellow pigment produced on agar media. Not acido-proteolytic, lactose not fermented.   *Micrococcus luteus.*

b. No pigment produced on agar media, not acido-proteolytic. Utilizes urea as a sole source of nitrogen. Acid from lactose.
Micrococcus ureae.
B. Pink, orange-red, or red pigment produced on agar media in young cultures.
1. Gelatin liquefied slowly. Produces rose-colored pigment.
Micrococcus roseus.
2. Gelatin not liquefied. Produces light, flesh-colored pigment on agar slants. *Micrococcus rubens.*

KEY TO THE SPECIES OF THE GENUS STAPHYLOCOCCUS

I. Ferments mannitol. Coagulase positive. *Staphylococcus aureus.*
II. Does not ferment mannitol. Coagulase negative.
Staphylococcus epidermidis.

KEY TO THE SPECIES OF THE GENUS GAFFKYA

I. Does not hydrolyze gelatin. *Gaffkya tetragena.*

KEY TO THE SPECIES OF THE GENUS SARCINA

I. Aerobic.
A. Urea not converted to ammonium carbonate (hydrolyzed).
1. Yellow pigment produced.
a. Litmus milk alkaline, coagulated. *Sarcina lutea.*
b. Litmus milk alkaline, not coagulated. *Sarcina flava.*
2. Orange pigment produced. *Sarcina aurantiaca.*

Family Neisseriaceae

Gram-negative cocci, frequently occurring in pairs. Nonmotile.

KEY TO THE SPECIES OF THE GENUS NEISSERIA

I. Grow well on ordinary laboratory media at 22°C.
A. No pigments produced.
1. No acid from any carbohydrate. *Neisseria catarrhalis.*
2. Acid from glucose, fructose, maltose, and sucrose.
Neisseria sicca.
B. Pigments produced.
1. Acid from fructose.
a. No acid from sucrose, yellow pigment. *Neisseria flava.*
b. Acid from sucrose, yellow pigment. *Neisseria perflava.*

IDENTIFICATION OF UNKNOWN BACTERIA

Family Pseudomonadaceae

Straight gram-negative rods; motile by means of polar flagella (single or tuft), or nonmotile. Often produce pigments, grow well on ordinary media but are generally weak fermenters.

KEY TO THE GENERA OF THE FAMILY PSEUDOMONADACEAE

I. Attack glucose fermentatively or aerobically.
   A. Many produce pigments.   Genus *Pseudomonas.*

KEY TO THE SPECIES OF THE GENUS PSEUDOMONAS

I. Produce diffusible pigments, usually yellow, green, or blue (ability to produce pigments may be lost).
   A. Grows in gelatin and liquefies it.
      1. Grows readily in 37° to 42°C. range.
         a. Litmus milk is alkaline. Indole usually not produced.
                                         Pseudomonas aeruginosa.
      2. Grows only poorly or not at all at 37°C.
         a. Litmus milk alkaline, indole is not produced.
                                         Pseudomonas fluorescens.

Short Key for the Family Determination of Unknown Bacteria

   A. Gram-negative rods, non-spore formers, often motile (peritrichous flagella).
      1. Active fermentation of carbohydrates with acid production and frequently gas.   Family: *Enterobacteriaceae*
      2. Poor fermentative qualities on carbohydrate media (usually ferment glucose only, with gas formation only: often no fermentation).   Family: *Achromobacteriaceae*
   B. Gram-negative rods, non-spore formers, often motile by single polar flagella.
      1. Poor fermentation of carbohydrates if any at all frequently forming pigments.   Family: *Pseudomonadaceae*
      2. Same as above but rods curved.   Family: *Spirillaceae*
   C. Gram-positive rods, non-spore formers, usually non-motile.
      1. Active fermentation of carbohydrates with acid but no gas. Some types are microaerophilic, catalase negative.
                                         Family: *Lactobacillaceae*
      2. Pleomorphic, frequently form metachromatic granules.
                                         Family: *Corynebacteriaceae*

D. Gram-positive rods, spore formers, usually motile.
   1. Good fermentation ability, good proteolysis and aerobic to facultative. Family: *Bacillaceae*
   2. Excellent fermentative ability with gas production, anaerobic. Family: *Clostridium*
E. Gram-negative cocci, non-motile, frequently in pairs. Family: *Neisseriaceae*
F. Gram-positive cocci, non-motile, non-spore formers.
   1. Some fermentative ability, sometimes proteolytic, cells arranged in indefinite patterns (except chains) from broth culture. Family: *Micrococcaceae*
   2. Excellent fermentative ability with acid but no gas, never proteolytic, usually arranged in pairs, short or long chains. Tend to grow on ordinary media only with difficulty. Family: *Lactobacillaceae*

SPECIFIC CHARACTERISTICS OF SELECTED BACTERIA
(See last page for key to symbols used)

	Gelatin Hydrolysis	Motility	Gram Stain	Litmus Milk Reaction	Nitrate Reduction	Glucose	Lactose	Sucrose	Maltose	Xylose	Inulin	Trehalose	Salicin	Dulcitol	Mannitol	Galactose	Arabinose	Glycerol	Dextrin	Starch	Glycogen	Rhamnose	Raffinose	Fructose	Sorbitol	Mannose	Indole	Voges-Proskauer Test	Methyl Red Test	Hydrogen Sulfide	Uric Acid	Citrate	Ammonium Phosphate	Casein Hydrolysis	Hemolysis	Capsules	Optimum Temperature (°C)	Urea Hydrolysis	Pigment	Ordinary Media Growth
Family Achromobacteriaceae																																								
1. Alcaligenes faecalis	−	+	−	2	V																						−	−	−	−					−	V	37	−		+
2. A. viscolactis	−	−	−	2	V																						−	−	−							V	20			+
3. A. metalcaligenes	−	−	−	2	V																						−	−	−	−							22			+
4. A. bookeri	+	+	−	3R	−																						−										37			+
5. A. recti	+	+	−	2	+																						−										37			+
6. A. marshallii	+	−	−	3	−																						−										30			
Family Bacillaceae																																								
1. Bacillus cereus	+	V	V	3	+	A	V	A	A	−	−		A		−	A	−	A	A	+		−	−					+				+				V	30			+
2. B. subtilis	+	V	+	3	+	A	V	A	A	A	A		(A)	−	A	A	A	A	A	+		(A)	(A)	A	(A)	A	−	+				+				V	30			+
3. B. polymyxa	+	+	V	5	+	(A)	(A)	(A)	(A)	(A)	(A)	(A)	(A)		(A)	(A)	(A)	(A)	(A)	V		(A)	(A)	(A)	(A)	(A)	−	+				−					30			
4. B. megaterium	+	V	V	3	−	A	V	A	A	A	A				A	A	A	A	A	+		−		A		A	−					+					30			
Family Corynebacteriaceae																																								
1. Corynebacterium xerosis	−	−	+	1	−	A	A	A	A							A								A			−									−	37	−		+
2. C. pseudodiphtheriticum	−	−	+	1	+	−		−	−																		−									−	37	+		+

IDENTIFICATION OF UNKNOWN BACTERIA

Family Enterobacteriaceae	Gelatin Hydrolysis	Motility	Gram Stain	Litmus Milk Reaction	Nitrate Reduction	Glucose	Lactose	Sucrose	Maltose	Xylose	Inulin	Trehalose	Salicin	Dulcitol	Mannitol	Galactose	Arabinose	Glycerol	Dextrin	Starch	Glycogen	Rhamnose	Raffinose	Fructose	Sorbitol	Mannose	Indole	Voges-Proskauer Test	Methyl Red Test	Hydrogen Sulfide	Uric Acid	Citrate	Ammonium Phosphate	Casein Hydrolysis	Hemolysis	Capsules	Optimum Temperature (°C)	Urea Hydrolysis	Pigment	Ordinary Media Growth	Lysine Decarboxylase
1. Escherichia coli	–	V	–	5R	+	(A)	(A)	V	(A)	(A)	–	–	V	V	(A)	(A)	V	V	–	–	–	(A)	V	(A)	(A)	–	+	–	+	–	–	–	+	–	V	–	37	–	–	+	+
2. E. aurescens	–	+	–	5R	+	(A)	(A)	–	(A)	(A)	–	(A)	(A)	–	(A)	(A)	(A)	(A)	–	–	–	–	–	(A)	(A)	(A)	+	–	+	–	–	–	–	–	–	–	37	–	+	+	–
3. E. freundii	V	V	–	6	+	(A)	(A)	V	(A)	(A)	–	–	V	V	(A)	(A)	(A)	(A)	–	–	–	(A)	(A)	(A)	(A)	(A)	V	–	+	+	–	+	–	–	–	–	37	–	–	+	–
4. Aerobacter aerogenes	V	–	–	5R	+	(A)	(A)	(A)	(A)	(A)	V	(A)	(A)	V	(A)	(A)	(A)	(A)	(A)	(A)	–	(A)	(A)	(A)	(A)	–	–	+	–	–	+	+	–	–	–	V	30	–	–	+	+
5. A. cloacae	V	+	–	5R	+	(A)	(A)	(A)	(A)	(A)	–	(A)	(A)	V	(A)	(A)	(A)	(A)	(A)	V	–	(A)	(A)	(A)	(A)	–	–	+	–	–	+	+	–	–	–	–	37	–	–	+	–
6. Klebsiella pneumoniae	–	–	–	6	+	(A)	(A)	A	A	A	–	–	A	–	(A)	(A)	A	A	–	–	–	–	A	A	A	A	–	+	–	–	–	+	–	–	–	+	37	V	–	+	+
7. Erwinia amylovora	+	+	–	4R	–	(A)	V	V	A	–	–	–	A	V	(A)	V	A	A	–	–	–	–	–	A	A	–	–	+	V	–	–	+	+	–	–	–	20	–	–	+	–
8. Serratia marcescens	+	+	–	4	+	(A)	–	V	A	–	–	–	–	–	(A)	(A)	–	–	–	–	–	–	–	–	–	–	–	+	–	–	–	+	–	–	–	–	30	–	+	+	–
9. S. indica	+	+	–	4	+	A	–	A	A	–	–	–	–	–	–	–	–	–	–	–	–	–	–	–	–	–	–	–	–	–	–	+	–	–	–	–	35	–	+	+	–
10. Proteus vulgaris	+	+	–	3	+	(A)	–	(A)	(A)	(A)	–	(A)	–	–	–	(A)	–	V	–	–	–	V	–	(A)	(A)	(A)	+	–	+	+	–	+	–	–	–	–	37	+	–	+	+
11. P. mirabilis	+	+	–	3	+	(A)	–	(A)	(A)	V	–	–	–	V	–	(A)	–	–	–	–	–	V	–	(A)	(A)	(A)	+	–	+	+	–	+	–	–	–	–	37	+	–	+	+
12. P. morganii	–	+	–	2	+	(A)	–	–	–	(A)	–	–	–	V	–	(A)	(A)	A	(A)	–	–	–	–	–	–	–	+	–	–	–	–	–	–	–	–	–	37	+	–	+	+
13. Salmonella choleraesius	–	+	–	2	+	(A)	–	–	(A)	V	–	–	–	V	(A)	(A)	V	–	–	–	–	V	–	–	(A)	–	–	+	+	+	–	+	–	–	–	–	37	–	–	+	+
14. Sal. typhimurium	–	+	–	2	+	(A)	–	–	(A)	(A)	–	V	–	V	(A)	(A)	(A)	–	–	–	–	–	–	–	A	(A)	–	–	+	+	–	V	–	–	–	–	37	–	–	+	+
15. Sal. typhosa	–	V	–	1	+	(A)	–	–	A	V	–	–	–	V	A	(A)	V	V	–	–	–	V	–	–	A	(A)	–	–	+	+	–	V	–	–	–	–	37	–	–	+	+
16. Sal. schottmuelleri	–	+	–	2	+	(A)	–	–	A	(A)	–	(A)	–	V	A	(A)	(A)	–	–	–	–	–	–	(A)	(A)	–	–	+	+	+	–	V	–	–	–	–	37	–	–	+	+
17. Shigella arabinotarda	–	–	–	2	+	A	–	V	A	–	–	–	–	V	A	A	A	–	–	–	–	V	A	A	A	–	V	–	+	–	–	–	–	–	–	–	37	–	–	+	–
18. Sh. flexneri	–	–	–	2	+	A	–	–	A	–	–	–	–	–	A	A	A	–	–	–	–	–	–	A	–	–	+	+	+	–	–	–	–	–	–	–	37	–	–	+	–
19. Sh. dispar	–	–	–	V	+	A	V	–	V	–	–	–	–	V	V	–	A	–	–	–	–	–	A	A	–	–	+	+	+	–	–	–	–	–	–	–	37	–	–	+	–
20. Sh. sonnei	–	–	–	4	+	A	A	A	A	–	–	–	–	–	A	A	A	–	–	–	–	A	A	A	–	–	+	+	–	–	–	–	–	–	–	–	37	–	–	+	–

116 IDENTIFICATION OF UNKNOWN BACTERIA

	Gelatin Hydrolysis	Motility	Gram Stain	Litmus Milk Reaction	Nitrate Reduction	Glucose	Lactose	Sucrose	Maltose	Xylose	Inulin	Trehalose	Salicin	Dulcitol	Mannitol	Galactose	Arabinose	Glycerol	Dextrin	Starch	Glycogen	Rhamnose	Raffinose	Fructose	Sorbitol	Mannose	Indole	Voges-Proskauer Test	Methyl Red Test	Hydrogen Sulfide	Uric Acid	Citrate	Ammonium Phosphate	Casein Hydrolysis	Hemolysis	Capsules	Optimum Temperature (°C)	Urea Hydrolysis	Pigment	Ordinary Media Growth
Family Lactobacillaceae																																								
1. Lactobacillus lactis	–	–	+	4	–	A	A	A	A	–	–	–	V		–	A	–	–	A	–		–	A	A	–	A											40			–
2. L. delbrueckii	V	–	+	1	–	A	–	A	A	–	–	–			–	A	–	–	A	–		–	–	A	–	A											45			–
3. L. casei	+	–	+	4	–	A	A	V	A	–	–	–	A		A	A	–			–		–	–	A	–	A											30			–
Family Micrococcaceae																																								
1. Micrococcus luteus	–	–	+	6	–	A	–	A	A	–	–	–			A			–		–							–						+				25		+	+
2. M. ureae	+	–	V	2R	–	A	A	A	A	–	–	–			A			A		–							–						+				25	+		
3. M. roseus	+	–	V	1-2	+	A	–	A	A	–	–	–			A			A		–													+				25		+	+
4. M. rubens	–	–	V	4	+	A	–	A	A	–	–	–			A			A		–																	37		+	
5. Staphylococcus aureus	+	–	+	4	+	A	A	A	A	–	–	A	–		A	V	–	A		–			–	A	–	V	–			+			–		B	–	37		+	+
6. Staph. epidermidis	+	–	+	6	+	A	A	A	A	–	–				A	V	–			–							–						–				37			
7. Gaffkya tetragena	–	–	+	6	–	A	V	A	A	–	–				A					–							–			–			–				37			+
8. Sarcina lutea	+	–	+	2	+	A	–	–	A	–					–												+			+							25		+	
9. S. flava	+	–	+	2	–	–	–	–	–																		–			–							35			
10. S. aurantiaca	+	–	+	3	–	–	–	–	–																		+			–							30		+	+

IDENTIFICATION OF UNKNOWN BACTERIA

	Gelatin Hydrolysis	Motility	Gram Stain	Litmus Milk Reaction	Nitrate Reduction	Glucose	Lactose	Sucrose	Maltose	Xylose	Inulin	Trehalose	Salicin	Dulcitol	Mannitol	Galactose	Arabinose	Glycerol	Dextrin	Starch	Glycogen	Rhamnose	Raffinose	Fructose	Sorbitol	Mannose	Indole	Voges-Proskauer Test	Methyl Red Test	Hydrogen Sulfide	Uric Acid	Citrate	Ammonium Phosphate	Casein Hydrolysis	Hemolysis	Capsules	Optimum Temperature (°C)	Urea Hydrolysis	Pigment	Ordinary Media Growth
Family Neisseriaceae																																								
1. Neisseria catarrhalis	–		–	1	+	–	–	–	–	–	–	–			–	–	–	–	–	–	–	–	–	–		–	–			–						–	37			+
2. N. sicca			–			A		A	A	–		–												A											V		37		+	+
3. N. flava			–			A		–	A															A													37		+	
4. N. perflava			–			A		A	A															A													37			
Family Pseudomonadaceae																																								
1. Pseudomonas aeruginosa	+	+	–	3R	+	–	–	–	–		–			–	–	–	–	–	–					–			–			+						–	37		+	+
2. Ps. fluorescens	+	+	–	2	+	A																					–										20		+	+

Key to the Fermentation Broth Results:

A = Acid
Ⓐ = Acid and gas
V = Variable
– = Negative or no change

Litmus Milk Code:

1 = No change
2 = Alkaline
3 = Alkaline with peptonization
4 = Acid, curd formation
5 = Acid and gas
6 = Acid, no curd
R = Signifies reduction

When an empty space is present, the information is insignificant or is not given in "Bergey's Manual of Determinative Bacteriology."

CHAPTER 7

ECOLOGY OF MICROORGANISMS

Ecology is the branch of biology that deals with the mutual relations among organisms and between them and their environment. In your laboratory work up to this point, emphasis has been given to study of the behavior of selected species of microorganisms as they grow in pure culture. Despite the scientific validity of gathering information based on pure culture techniques, from a purely biological point of view this can be misleading if not supplemented in other ways. In the definition of ecology, special emphasis should be directed to the phrase "mutual relations." This phrase implies that when microorganisms grow, live, and die in nature, a definite association effect is brought about, because the microorganisms must live in close proximity with other microorganisms. It is one thing to study the growth capabilities of a single pure culture of microorganisms in the relative protection of the test tube, and quite another thing to observe the same organism competing in nature with many other species for the same supply of food. Despite the fact that mutual association is often very likely to be an antagonistic or competitive one, there are many unusual and interesting beneficial associations that have developed during the evolution of modern microorganisms.

Several distinctly different types of *symbiosis* ("the living together in more or less intimate association of two dissimilar organisms") have been observed. Some of these symbiotic variations will be studied experimentally in this chapter. In most of these experiments the emphasis is directed to the intimate association between two or more organisms and the resulting effects of this association. In some of the experiments you will be inoculating more than one species of bacterium into one test tube. This is a distinct change from the previous experimental pattern. In the last part of the chapter, experiments are included that deal with a reasonably natural approach to symbiotic relationships in their customary surroundings.

In the first experiment, *commensalism* ("the living together of two different species in which one member gains benefit and the other is unaffected") is illustrated. The two species selected for this experiment are *Staphylococcus aureus* and *Clostridium sporogenes*. *Staphylococcus aureus* is a facultative organism and *Clostridium sporogenes* is a strict anaerobe. In this experimental situation, the gas tension requirement is the principal item of focus. These organisms differ in their requirement for oxygen; when the two are grown together in an environment that contains oxygen, development of a commensalistic situation occurs. Technically speaking, the *Cl. sporogenes* should be unable to grow in the tubes, since some oxygen is present in the medium. On the other hand, staphylococci are not only capable of utilizing oxygen but do so preferentially if it is available. Soon after inoculation, the staphylococci gradually create conditions conducive to the eventual growth of clostridia. Each organism is capable of growing in nutrient broth if proper oxygen tension is provided.

EXPERIMENT 7-1

Bacterial Commensalism

MATERIALS

1. Broth cultures of Staphylococcus aureus and Clostridium sporogenes (the latter grown in thioglycollate broth).
2. Nutrient broth (3).

PROCEDURE: Set up a series of three tubes of nutrient broth. Inoculate tube 1 with two loopfuls of Staph. aureus, and tube 2 with the same amount of Cl. sporogenes, making sure in the later case that the loop is immersed deep in the thioglycollate broth to obtain the anaerobes. Inoculate tube 3 with each organism, using the same amounts (resulting in a mixture of the two organisms in the same tube). Incubate the tubes at 37°C. for one to two days and examine them for turbidity as an indication of growth. To make sure that any turbidity corresponds to the organism inoculated and is not caused by contamination, make Gram stains of each tube. Record your results showing (a) number of the tube, (b) plus or minus turbidity (growth), and (c) the Gram stain picture for each tube.

In the next experiments, the phenomenon of *antibiosis* is illustrated in two different ways. Antibiosis is the association in which one or more species are harmed as a result of growing in close relationship with another species, because of production by the latter species of chemical secretions inhibitory to the growth of adjacent organisms. In this way competing species are removed from the competition for food. This phenomenon is widespread in nature, since there are many species that have developed the ability to produce antibiotics. You will first study a purely laboratory situation involving an organism with proven ability to produce an antibiotic; you will study its effect upon species growing in close association with it. In the second experiment you will set up the laboratory duplication of a natural situation. Diluted soil samples (teeming with all kinds of microorganisms) will be grown dispersed in nutrient agar Petri plates, and observations will be made for the presence of typical antibiotic-producing organisms. In both experiments, it is the *zone of inhibition* that tells if an antibiotic is being produced. Since these zones may be quite small (and sometimes irregular in shape), it is helpful to examine these plates with the aid of a magnifying device such as a colony counter or stereoscopic microscope using low power.

EXPERIMENT 7-2

Experimental Antibiosis

MATERIALS

1. Broth cultures of Micrococcus luteus, Escherichia coli, and an antibiotic-producing strain of Bacillus subtilis.
2. Nutrient agar talls (2).
3. Sterile Petri dishes (2).
4. Sterile 1 ml. pipets (2).

PROCEDURE: Melt two nutrient agar talls and allow them to cool to about 50°C. Inoculate tube 1 with 0.1 ml. of the M. luteus culture, using a sterile pipet. Be sure to observe aseptic technique. In tube 2 repeat the procedure but use the culture of E. coli this time. Immediately pour the two talls into sterile Petri dishes, and allow them to harden after gently swirling the contents to mix the organism with the agar (this may be done by rotating the tubes before pouring, if desired). The first agar plate is now thoroughly impregnated with M. luteus and the second plate with E. coli. Draw a single line of growth of B. subtilis over the center of each of the plates — that is, bisect the plate down the center with a line of growth. As this organism grows, it secretes an antibiotic into the medium. Observe the plates for several days for any evidence of this activity. Record your results and conclusions.

EXPERIMENT 7-3

Antibiosis in Soil

MATERIALS

1. Sample of rich soil.
2. Nutrient agar talls (5).
3. Sabouraud agar talls (3).
4. Sterile Petri dishes (6).
5. Pan balance.
6. 1000 ml. Erlenmeyer flasks (4 or 5 should be adequate for the entire class).
7. Nonsterile empty test tubes (3).
8. Broth cultures of Escherichia coli and Micrococcus luteus.

PROCEDURE: Using the pan balance, weigh out 2 g. of soil rich in organic matter and suspend in 1000 ml. of tap water in an Erlenmeyer flask. Stopper and shake well to mix thoroughly. This is a 1:500 dilution of the original soil sample. Remove 5 ml. of this suspension into a nonsterile test tube (none of these dilutions requires sterile materials) containing 5 ml. of tap water. Label this tube 1:1000 dilution. Next remove 5 ml. of your 1:1000 suspension into another test tube containing 5 ml. of tap water and label this latter suspension 1:2000. You now have three different dilutions of the original soil sample — 1:500, 1:1000, and 1:2000. Melt an agar tall each of nutrient agar and Sabouraud agar for each of the three dilutions and cool to about 40°C. Pipet 1 ml. of each dilution into each of the melted talls — that is, 1.0 ml. of the 1:500 dilution into the nutrient agar and 1.0 ml. of the same dilution into the Sabouraud agar. Repeat with the next higher dilution, and so on. Once the dilution of soil is pipetted into the melted agar, rotate the mixture quickly with the hands and pour into a properly labeled Petri dish. Continue until all the talls are poured. After two or three days incubation in your drawer, observe for evidence of antibiosis by zones of inhibition. When a zone of inhibition is found, isolate the colony producing the antibiotic into a stock culture (if none is apparent, repeat using a different soil sample until you are successful). It may be necessary to streak for isolation to make sure your stock culture is pure. Next, experimentally establish a rough "spectrum" for your antibiotic in terms of gram staining characteristics, i.e., is it effective against gram-negative (use E. coli culture provided) or gram-positive (use M. luteus culture provided) or both? Use the procedure employed in the previous experiment on antibiosis. Be sure, of course, to record what type of an organism, in general, you have (gram stain, bacteria, yeast, mold, etc., plus a few other outstanding characteristics that you may observe). Also in your observations be sure to note the different kinds of growth that appeared on the two kinds of agar used, even though the inoculum was the same for each. What kind of organisms tend to predominate on each. Why is this so?

The next experiment demonstrates the type of association wherein two species produce a change that neither would be capable of doing alone. This phenomenon is called *synergism*. The particular change you will look for involves an *end product of metabolism*. The cultures in this experiment should be incubated for several days if synergism is not apparent at first. It sometimes takes longer to develop. The first two tubes in each set-up represent the metabolic changes brought about by the individual organisms, so in your observations you should compare the results obtained with and without the association.

EXPERIMENT 7-4

Bacterial Synergism

MATERIALS

1. Broth culture of Staphylococcus aureus, Proteus vulgaris, and Escherichia coli.
2. Lactose fermentation broth (3).
3. Sucrose fermentation broth (3).

PROCEDURE: Set up three tubes of lactose fermentation broth, numbering them 1, 2, and 3. Inoculate tube 1 with Staph. aureus, tube 2 with P. vulgaris, and tube 3 with both organisms. Repeat the procedure with sucrose broth, but this time tube 1 receives Staph. aureus, tube 2 receives E. coli, and tube 3 receives both organisms. Incubate the tubes at 37°C. for one to five days and record the results as acid, acid and gas, or no change.

The next experiments illustrate the normal flora of selected parts of the human body. The "normal flora" concept is an important part of ecology, because the conditions that determine which species shall predominate in any given location are determined by the biological interaction between contiguous species as well as conditions of the physical and chemical environment. Despite equal opportunity to grow in all locations, the predominating species differ with the different locations. This section gives the student some excellent opportunities to study and interpret his observations on the real meaning of ecology. In your analysis, attempt to decide why each situation favors a certain species, and why such a situation is not necessarily permanent. The specific sites selected include the normal human skin, throat, and teeth. Included in these experiments is a test that has some merit in the field of dentistry, the so-called Snyder test. Salivary samples are analyzed for the presence of species of *Lactobacillus,* particularly for *L. acidophilus.* It is generally agreed that dental caries, or decay, is a result of the acid produced during fermentation of bits of carbohydrate on the tooth surfaces by these organisms. The enamel of the tooth in the immediate vicinity is slowly decalcified and softened as a result of the continued presence of the organic acids that represent the end-products of fermentation by these organisms. Decay is thereby initiated.

The Snyder test employs an agar with an adjusted pH of about 4.8, which is *too low for most species* but is suitable for growth of the *Lactobacilli.* An indicator in the agar determines if further acidification (more growth) is taking place by turning from green to yellow, and this change in color is interpreted as relating to the number of *Lactobacilli* originally present in the saliva sample — that is, the more acid produced (and therefore, the more yellow the agar appears to be), the more of these organisms originally present.

One of the most fascinating examples of symbiosis in nature concerns the interaction between members of the bacterial genus Rhizobium and the group of plants generally known as legumes (peas, alfalfa, clover, etc.). These soil bacteria are able to invade the root tissue of the appropriate plant types and establish themselves in what would appear at first to be a genuine infection, i.e., the bacteria multiply at the nutritional expense of the host. The plant initially responds in a typical fashion: a tumor-like mass of proliferating root cells is formed around the developing bacterial colony as a means, presumably, of circumscribing the growth of the microorganisms. However the bacteria, in this environment begin to convert ("fix") atmospheric N_2 into nitrogen compounds that can be utilized by the plant as nutrient. At that point, an "infection" becomes a mutualistic association in which both members benefit from the arrangement.

The biochemical events which make all of this possible have yet to be worked out in detail, but they will ultimately prove to be a fascinating story, based on what we already know. For example, the Rhizobia undergo a remarkable morphological transformation when moving from soil to plant, in that their shape is very inconstant within the plant as compared to the way they look outside of it. In addition, the Rhizobia apparently are unable to fix atmospheric nitrogen outside the plant, requiring the association with legumes in order to do this. Another interesting facet of the symbiotic arrangement is the fact that healthy root nodules are of pink or reddish color inside, due to the production of hemoglobin or a molecule very similar to it — a substance that neither legume or bacterium produces alone. Undoubtedly this substance plays a vital role in the biochemistry of nitrogen fixation, yet to be elaborated.

The experiment to follow is primarily designed to demonstrate the changes in bacterial morphology in this symbiotic relationship.

EXPERIMENT 7-5

Symbiotic Nitrogen Fixing Bacteria

MATERIALS

1. Suitable plants (clover, peas, etc.) recently removed from soil, kept moist.
2. Tall of yeast-extract mannitol agar (1).
3. Loeffler's methylene blue.
4. Razor blades (1).
5. Sterile Petri dish (1).
6. Tube of sterile tap water (1).

PROCEDURE: Remove a large nodule from the roots of the legume provided and rinse it off thoroughly in tap water to remove the adherent dirt particles. Melt a tall of yeast-extract mannitol agar, cool to 45°C. and pour into a Petri plate. When the plate has hardened, put the cleaned nodule on a microscope slide and cut in half with a razor blade. Observe the color of the nodule interior. Next crush both halves of the nodule in a drop of sterile tap water with the top of your inoculating loop handle until the water becomes quite turbid. Streak a loopful of this suspension for colony isolation on the agar plate, invert and incubate at room temperature for about a week.

Remove the large debris from the microscope slide and spread the suspension on your slide to about the size of a quarter, air-dry and fix gently with heat. Flood the smear with methylene blue for about 4-5 minutes, rinse, dry and examine with your oil-immersion lens. Note the pleomorphic forms and make a drawing of representative bacteria. Save this slide to compare with the slide noted below.

After about a week, observe your streak plate for isolated colonies that show a pronounced glistening appearance with a tendency to become white at the edges. These are Rhizobia colonies. Remove a bit of growth from an isolated colony of this appearance and prepare a smear of it, followed by a gram stain. Observe under oil-immersion and make a drawing of representative forms. Compare these organisms with those you stained with methylene blue above.

EXPERIMENT 7-6

The Relationship of Microorganisms to Dental Decay

MATERIALS

1. Chewing gum or paraffin.
2. Trypsin-digest agar short containing bromcresol green indicator (Snyder test agar) (2).
3. Sterile Petri dish (1).
4. 1.0 ml. pipet (1).

PROCEDURE: Chew a piece of gum or paraffin for several minutes to stimulate salivary flow, then deposit a saliva sample in a sterile Petri dish. Pipet 0.2 ml. of this saliva into a melted and cooled tube of Snyder test agar (short). Be sure the agar is not too hot. Roll the tube between the palms of the hands to distribute the saliva evenly throughout the agar. Allow to harden in the upright position, and incubate at 37°C. Make daily observations for the next three days. Since this three-day observation period is absolutely necessary, make sure that you do not begin this exercise after Tuesday. When you make your observations, you should always compare your inoculated tube to an uninoculated tube, using reflected light and holding both tubes against a white background. The control should be bluish-green. In recording results the following code should be used:

Positive: Change in color so that green is no longer dominant; should be recorded as 2+ to 4+, depending on the degree of change.
Negative: No change in color or only slight deviation, but green still dominant; is recorded as 0 or 1+.
Interpretation: Caries activity (relative number of Lactobacilli):

Hours Incubation

	24	48	72
Marked:	positive	———	———
Moderate:	negative	positive	———
Slight:	negative	negative	positive
Negative	negative	negative	negative

Note: The data indicate only what is happening at the time the specimen is collected and does not indicate past history. At least two different specimens should be collected within a two-week period to compare for accuracy. Record your results from the two separate trials, and in your observations discuss the relation between the results obtained from this test and your knowledge of your dental state of health.

It is important to note that a layer of oil often prevents efficient removal of skin bacteria. Therefore it is helpful to wash the skin first with soap and water to assist in making microorganisms on the skin more available to swab removal. The next experiment would take much longer and be more complicated if a thorough study were undertaken. Many organisms present may not grow well (or at all) on the surface of nutrient agar. Perhaps the nutrient is inadequate or the oxygen tension too high. You will observe principally aerobes or facultative organisms that are relatively nonexacting in their nutritional requirements, since these organisms grow in nutrient agar that is not especially rich. The pH and osmotic pressure on the agar are different from the pH and osmotic pressure on the surface of the skin. Think about some of these facts and they will take on a great deal more significance as you investigate this problem.

EXPERIMENT 7-7

Normal Flora of the Human Skin

MATERIALS

1. Sterile swabs (2).
2. Sterile water blanks (2).
3. Nutrient agar for plating.
4. Sterile Petri dishes (2).
5. Glucose, sucrose, and lactose fermentation tubes (2 each).
6. Peptone-iron agar short or T.S.I. agar short (2).

PROCEDURE: In this exercise, you are interested in dislodging microorganisms that normally live on the skin. Using a sterile swab and a sterile distilled water blank, swab two different areas of the skin that have been washed with soap and water. Deposit a sample from each area upon a separate Petri dish that has been poured with nutrient agar. Be sure to use only one swab per selected area, and premoisten the swab with sterile distilled water for each area. After depositing the sample of microorganisms on a small area of the plate, streak away from the deposited material with a sterilized inoculating loop in order to isolate as many kinds of organisms as you can. Repeat this procedure with the other sample. Determine after suitable incubation at 37°C. which type(s) tend to predominate. You can determine this roughly by observing the single most frequently occurring colony type on each plate, and more precisely by characterizing individual colonies as to gram stain, motility, selected fermentations (glucose, sucrose, and lactose), spore formation (use simple methylene blue stain and look for unstained spores), and hydrogen sulfide production. On the basis of these observations you should be able to partially identify the predominating species. Be sure to mention everything you did and found in your observations.

Assuming that you are in good health, there should be an interesting variety of normally occurring microorganisms in your throat (including several pathogenic varieties). The next experiment will be performed with the same idea in mind as with the experiment on the normal flora of the skin. You will name or tentatively identify the predominating species isolated. Since many of the commonly occurring throat microorganisms require an enriched diet, nutrient agar with 5 per cent whole blood added is used.

EXPERIMENT 7-8

Normal Flora of the Human Throat

MATERIALS

1. Blood agar plate (1).
2. Sterile swab (1).
3. Glucose, lactose, sucrose, mannitol, trehalose, and inulin fermentation tubes (3).

PROCEDURE: This procedure is much the same as in experiment 7-7 except there is no need to wet the swab before using. The instructor will demonstrate the proper technique for taking a throat sample. Deposit the material from the swab onto the surface of your blood agar plate and streak carefully for isolation. Incubate for 24 to 48 hours at 37°C. Look for the most commonly occurring colony types. Select three of these for further study and partial identification. You may also notice a zone of clearing around some colonies. This is beta hemolysis, caused by the secretion of an exoenzyme by the organisms within the colony. You may see colonies that are surrounded by a zone of greenish discoloration. This is called alpha hemolysis, also caused by action of an exoenzyme. Be sure to gram stain and generally characterize any hemolytic colonies of either type. The fermentable products inulin and trehalose should be used when identifying a gram-positive coccus, since these substances assist in distinguishing between the various streptococci and Diplococcus pneumoniae, all of which are frequent inhabitants of the human throat. Streptococci ferment trehalose and the pneumococci ferment inulin. Staphylococcus aureus may also be present, and the fermentation of mannitol accompanied by a typical clustering and gram stain reaction will help you to partially confirm the presence of this organism. Staphylococcus aureus is a frequent inhabitant of the normal flora of the human throat in the absence of apparent infection. Make your write-up similar to that of the preceding exercise.

Note: If anyone in the class has a sore throat when this experiment is being performed, the instructor should be notified. He will set up a demonstration plate from the lesions to show the difference between flora from the normal throat and flora from an abnormal throat.

CHAPTER 8

APPLIED MICROBIOLOGY

Man has learned to use his knowledge of the various microorganisms to safeguard his health, to manufacture various foods and to conduct other activities generally beneficial to him. These activities are often categorized as the applied aspects of microbiology, which is contrasted to the research devoted primarily to expanding our knowledge of the field. The two aspects of the subject are closely interwoven, since in many ways they depend on each other for sustenance. A pure science must ultimately be a dead end if it has no practical application, and the latter cannot exist without the knowledge generated by the former. In this chapter we will examine some of the practical or applied aspects of microbiology to illustrate the role of these activities in all our lives.

It is undoubtedly obvious to all of us that water is fundamental to all life. Thus man has spent huge sums of money and much energy providing for the distribution of this vital substance to meet his needs. Not so obvious are the microbiological problems inherent in the process. Water proved to be an excellent vehicle for the transporting of pathogenic organisms, particularly those associated with improperly treated or untreated sewage. Typhoid fever and the various bacterial dysenteries are prominent in this group. The spread of these diseases gave rise to such counteracting procedures as chlorination of water supplies.

There are always situations arising even today where one must test water to determine if it is safe for human consumption. The bacteriological analysis of water which you will perform below is a standard procedure derived to ascertain whether a water sample is free of possible sewage contamination, and therefore, indirectly, whether or not the possibility exists that it contains typhoid or dysentery-causing bacteria. The assumption is made that any water showing evidence of sewage contamination is unsafe for human consumption.

Ideally one would achieve this goal by showing the presence of the pathogens themselves, but this has proved to be impractical. They are usually diluted out to such an extent that the comparatively small water sample used for testing would often prove to be free of the pathogens. Therefore, water analysis of this type looks for organisms that are far more numerous in any sewage sample, those of the so-called *coli-aerogenes* group of the family Enterobacteriaceae. In the final analysis, it is *Escherichia coli* which is being tested for, since it is not necessarily an index of fecal pollution to find *Aerobacter aerogenes*. *A. aerogenes* is known to be widespread in its distribution (other than sewage) and could therefore give a possible false indication of sewage contamination.

The standard test is divided into three subdivisions, known as the *presumptive, confirmed* and *completed* tests. The basis for each of the tests is to narrow the identification more precisely to the bacterium *E. coli* or to rule it out. Advantage is taken of the fact that very few organisms will ferment lactose with the production of acid and gas but many of the coliform bacteria will. Summarily, the tests are designed to isolate gram-negative, non-spore forming rods that ferment lactose with acid and gas production.

EXPERIMENT 8-1

The Bacteriological Analysis of Water

A. Presumptive Test

MATERIALS

1. Durham fermentation tubes of lactose broth (3).*
2. Several number-coded water samples, some positive for coliform bacteria and some negative.

PROCEDURE: Each student should test three unknown water samples. Using the usual aseptic precautions, pipette 1 ml. of each of your assigned water samples into three separate tubes of lactose broth marked with the appropriate sample code. Incubate all three tubes at 37°C. for 24 hours. Observe your tubes for gas at the end of the 24-hour period. If any tubes show gas production, it is considered to be a positive presumptive test. Save any positive tubes for use in the confirmed test. Continue incubation of any negative tubes for another 24 hours. The appearance of gas during the second 24-hour interval is considered to be a doubtful test. If there is no gas present after 48 hours, the sample is considered to be negative and therefore safe for consumption.

---

*For this experiment, regular-sized tubes of fermentation broth will be used, although the usual procedure involves using 10 ml. double-strength tubes inoculated with 10 ml. of water sample.

B. Confirmed Test

MATERIALS

1. Eosin-methylene-blue (E.M.B.) agar talls (1-2).
2. Sterile Petri plates (1-2).
3. Any tubes showing gas from the presumptive test.

PROCEDURE: The production of gas in lactose broth does not necessarily mean that coliform bacteria are growing in the medium. There are various false positive situations that could occur and these must be ruled out. Melt your E.M.B. agar tall or talls (depending on the number of presumptive positive tubes you obtained in the previous test), allow to cool to about 45°C., then pour into sterile Petri plates and allow each to solidify. Label each plate with a code number corresponding to a positive lactose fermentation tube. Streak growth from lactose broth tubes onto appropriately labeled E.M.B. plates, being careful so as to achieve isolated colonies. Incubate all plates inverted at 37°C. for 48 hours. After the incubation period, observe each plate for the presence of typical coliform colonies. E. coli will usually produce dark colonies with a characteristic greenish metallic sheen on the top. Other coliforms will produce pigmented colonies (lactose fermentation) of varying types. Any pigmented colony should be considered as positive for coliform bacteria, and these are to be saved for the completed test. E.M.B. agar is generally selective for gram-negative bacteria, so any lactose fermentation originally observed in the presumptive test that might have been due to gram-positive or anaerobic bacteria is generally eliminated from contention by this agar.

C. Completed Test

MATERIALS

1. Nutrient agar slants (3-4).
2. Lactose broth (3-4).
3. Tryptone broth (3-4).
4. MR-VP broth (3-4).
5. Sodium citrate broth or Simmon's citrate agar slant (3-4).

PROCEDURE: The purpose of the completed test is to specifically identify the lactose-fermenting organisms isolated on E.M.B. agar in the previous experiment. Select 1 or 2 seemingly typical colonies of E. coli and 1 or 2 colonies that are lactose-positive but lacking the green sheen on the surface. Do this for each positive water sample. Transfer some growth from each colony to a nutrient agar slant and to a lactose fermentation broth. Incubate for 24-48 hours at 37°C. Observe the lactose broth for acid and gas. Prepare gram stains from your slants and observe for the presence of small gram-negative rods.

The final steps of the completed test are for the purpose of separating E. coli from A. aerogenes. This can be done by a combination of tests known as the IMVIC tests (Indole, Methyl Red, Voges-Proskauer and Citrate utilization). Inoculate each of these tests with growth from typical lactose-fermenting colonies of gram-negative rods which you have isolated above. From these tests you should be able to positively identify the organisms as E. coli, A. aerogenes or both. Record all of your results. What do you conclude regarding the water sample you analyzed? As an item of added interest, students who have access to "raw" (untreated) well water should be encouraged to bring samples to class for parallel analyses with your assigned ones.

In the usual sense of the word, growth means increase in size of an organism, but in the unicellular world growth means simply an increase in numbers. In a real sense, measurement of bacterial growth is a study of population changes. Thus a bacterial population is observed to go through several "phases" of growth during the course of a 24-hour incubation period (although some take much longer than this time interval to pass through all the phases). The basic data needed for such a study is simply the number of viable organisms as a function of the time of incubation. This would be fairly simple to achieve when dealing with larger organisms, but poses special problems when one needs to count bacteria, since they are so small and numerous.

Although technically speaking the counting of bacteria is not a usual part of the subject of applied microbiology, it is used extensively in the latter and it will be included therefore in this section. The experiment to follow will be indicative of the general census-taking procedure employed in all situations requiring it. The procedure is a difficult one to perform accurately and individual counts may vary widely, but in the hands of a skilled operator it can be surprisingly reproducible.

The most obvious source of error to be guarded against is in volumetric measurements. When you are pipeting 1 ml. of a bacterial culture containing 1×10^8 organisms per ml., even the difference of a half drop (about 0.025 ml.) in successive measurements can add up to enormous errors. That volume would contain 250,000 organisms and thus could make a large difference in the final calculations. Therefore, when you are pipeting a specified volume, make sure you measure the volume as precisely as possible. Hurried measurements are almost guaranteed to give you erratic results.

Another major source of error is the tendency for bacteria to clump or adhere to each other in small clusters. Since the assumption is made that each colony appearing in the Petri plates is derived from a single original bacterium, any colonies actually derived from two or more adherent organisms is going to be reflected in inaccurate counts, particularly in the higher dilutions. Thorough mixing and shaking of the bacterial suspensions is performed to minimize this kind of error.

Counting errors are more apt to be minimized by averaging together several counts taken at the same time interval. The various groups should be encouraged to take their original samples within a few minutes of each other and their ultimate counts averaged together before graphing the results.

EXPERIMENT 8-2

The Bacterial Growth Curve

MATERIALS

1. Erlenmeyer flask containing 100 ml. of brain-heart infusion inoculated with E. coli 10-12 hrs. earlier and incubated at 37°C. (1 for every 8 students)*
2. Sterile water blanks, 99 ml. and 9.9 ml. (3 and 6).
3. Nutrient agar talls (8).
4. Sterile pipets, 1 ml. (15).
5. Sterile Petri plates (6).
6. Colony counter (optional).
7. Spectrophotometer (optional) and cuvettes.

PROCEDURE: An Erlenmeyer flask has been previously inoculated for you with a culture of E. coli in brain-heart infusion, to serve the needs of the entire class for this experiment. Mark your sterile water blanks with the elapsed time for each of three intervals, i.e., 0 hr. − 1 hr. − 2 hr. or whatever time intervals are suggested to you by your instructor. At each time interval you are to procure exactly 1 ml. of the growing culture and transfer it to a 99-ml. water blank. Expel the contents of your pipette and aspirate from the blank several times to wash out the contents thoroughly. Cap the blank tightly and shake well to break up any clumps of bacteria. Now pipette exactly 0.1 ml. of this 1:100 dilution of the original 1 ml. of bacterial culture into a 9.9 ml. water blank, repeating the pipette-rinsing procedure as before. After closing the test tube, swirl the contents well by snapping the bottom of the tube. Repeat the procedure again by transferring exactly 0.1 ml. of this 1:10,000 dilution into another 9.9-ml. water blank. After thorough mixing the last tube will represent a 1:1,000,000 dilution of the original 1 ml. of culture. Pipet exactly 1 ml. of the last 2 dilutions into tubes of nutrient agar that have been melted and cooled to 45°C., swirl the contents well to mix and pour into Petri plates marked 1:100,000 and 1:10,000,000 or 1:10^5 and 1:10^7. Allow the agar to harden, invert and incubate for 24-48 hours at 37°C.

---

*Ideally the culture should be aerated or swirled constantly during incubation to achieve maximum growth rate.

APPLIED MICROBIOLOGY 133

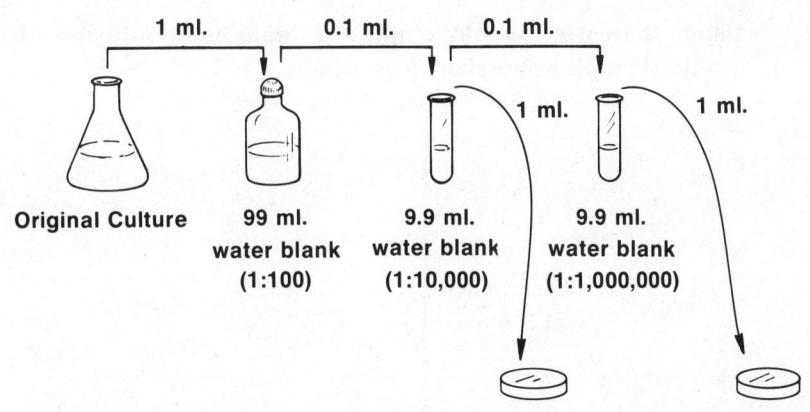

Repeat this whole procedure for the second hour (or whatever time interval chosen) and again for the third. Be sure to label your Petri plates appropriately as to time interval and dilution.

After incubation is complete, count the colonies growing in and on the agar in each Petri dish. Do not count plates whose colony count exceeds 300 or is less than 30. Calculate the population of bacteria per ml. of original culture by multiplying the plate count times the dilution factor. For example, a plate count of 39 in a Petri plate of 10^5 dilution would be equal to 39×10^5 or 3,900,000 bacteria/ml. Plot your data on graph paper as follows:

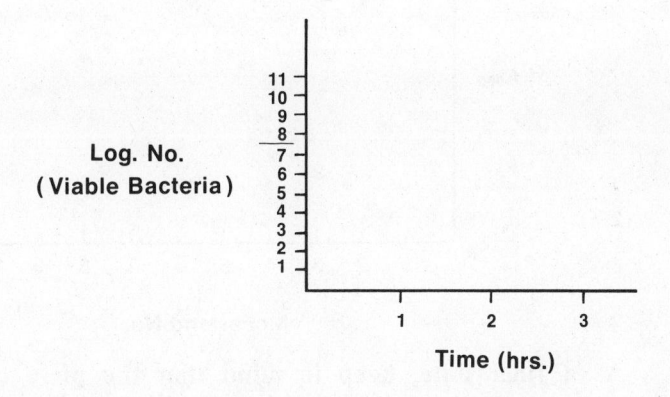

Ideally you would do population counts over a much longer time interval, at least 12-18 hours, but this is generally impractical in a student laboratory unless later lab sections can be used to continue the count. The data you collect can be combined with that obtained by others in the class and average values calculated. If time were no factor, you could continue sampling throughout a 24-hr. interval and ultimately reconstruct the total bacterial growth curve.

Optional. If a spectrophotometer such as the Bausch & Lomb Spectronic-20 is available, readings could be taken directly on the original culture with the machine being used as a turbidometer. Set the wavelength control to 600 nm. and blank the machine with samples of uninoculated brain-heart infusion broth. All readings taken thereafter would employ a brain-heart infusion blank. You would then plot absorbance vs. time:

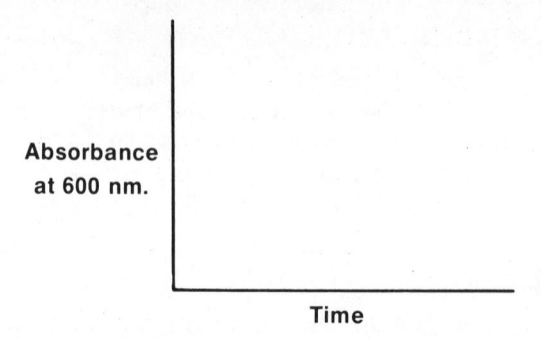

[Graph: Absorbance at 600 nm. vs. Time]

If you read the turbidity associated with the McFarland scale (p. 85), these values obtained on your growing culture can be converted into reasonable approximations of actual bacterial numbers. In order to do this, you would have to construct a quantitation curve from your McFarland tubes as follows:

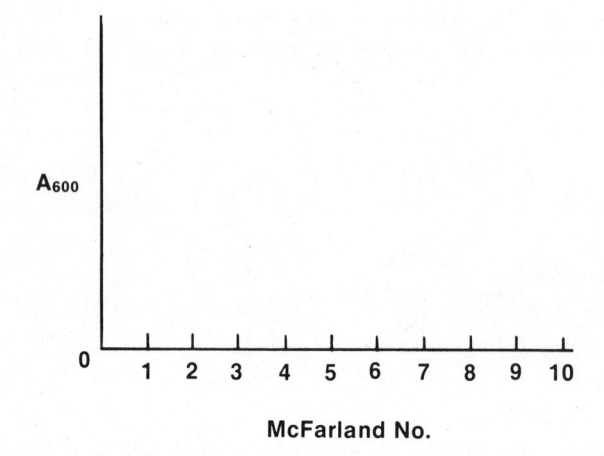

[Graph: A_{600} vs. McFarland No. (0 through 10)]

As a final note, keep in mind that the plate count method enumerates viable (but not dead) bacteria, whereas the spectrophotometer and McFarland methods would quantitate all bacteria, dead and alive. Since there are very few dead bacteria in the culture until nearing the end of the phase of logarithmic increase, the latter devices will be satisfactory until that point. However, as increasing numbers of dead organisms accumulate during the crisis phase, these devices are rendered inaccurate. Thus a viable count is necessary in order to observe the changes that take place in the bacterial population as it passes through its cycle.

One of the most useful applications of man's microbiological knowledge has been that in which he has concerned himself with the prevention of the spread of infectious diseases by means of the various foods that are particularly suited to this process. Among the foods which are prime suspects in spreading disease bacteria, milk is certainly one of the best known. There are several reasons why this situation prevails. As the student should be well aware by now, milk is a good medium for the growth of many different microbial types. Additionally, there are many opportunities for the inoculation of milk with microorganisms not normally associated with it in the processes of procurement and distribution. Finally, there is the possibility that the cow from whence the milk was derived was infected with a disease that is also communicable to man.

In order to render milk safe for human consumption, the well-known pasteurization procedure is applied to most commercially sold milk. This process consists of raising the temperature of the milk to 63°C. and holding it there for 30 minutes. Alternatively, a "flash" pasteurization may be effected by heating the milk to 72°C. for a 30-second period. The purpose of the procedure is to make sure that all *Mycobacterium tuberculosis* that might be present are killed, since this pathogen is the most resistant to heat of the various pathogenic bacteria that are commonly found in milk (there are no spore-forming bacteria of any medical concern found in milk as a rule, and thus this procedure will not kill this kind of bacterium). The pasteurization of milk is designed therefore to render it safe for consumption without destroying its palatability (many people who were raised on raw milk find the taste of pasteurized milk much less desirable). Ordinarily the testing of pasteurized milk by the Public Health department entails the use of a procedure called the Phosphatase Test. Milk normally contains a phosphate-hydrolyzing enzyme called phosphatase which is inactivated by heat (of pasteurization). If any of this enzyme remains after pasteurization, it indicates that the process was faulty (although there are other reasons which can be back-checked).

There are many non-pathogenic bacteria normally found in milk to a greater or lesser extent, even after pasteurization. Although they are not a disease problem, they can bring about changes in milk, some of which will make it unpalatable. Common genera to be found normally in milk include *Micrococcus, Pseudomonas, Alcaligenes, Lactobacillus* and others. In general, the type to gain ascendency in the milk will depend upon the conditions under which it is held, particularly the temperature. In general, cooler holding temperatures encourage the growth of fermentative varieties which are responsible for the normal souring of milk, while warmer temperatures will encourage the growth of the putrefactive type of organisms which give rise to disagreeable odors, etc.

In the experiments to follow, you will examine some of the properties of milk from a bacteriological point of view. Initially you will perform a direct plate count for the purpose of enumerating bacteria in several milk samples. In general, the plate count is an excellent indicator of the quality of the milk sample. It does not distinguish one type of organism from another, but it does serve to indicate the general sanitation and care that has been invested in the prior handling of the milk. The more organisms found in such a count, the greater the probability that there are human pathogens present. Milk which has not been pasteurized (raw milk) will naturally have a higher count than that which has been, but this does not mean that the raw milk is any lesser in quality. On the contrary, because of the more rigid requirements for the production of raw milk, it is usually of better average quality than the pasteurized variety. Nevertheless it is useful to examine the two types together in the direct plate count, as will be done below.

EXPERIMENT 8-3

The Plate Count of Bacteria in Milk

MATERIALS

1. Milk samples, raw and pasteurized.
2. Standard Methods Agar (Tryptone Glucose Yeast Agar) talls (3-4).
3. 1 ml. pipettes (6-8).
4. Sterile 9 ml. water blanks (3-4).
5. Sterile Petri dishes (3-4).
6. Colony counter.

PROCEDURE: For this experiment, the class should be divided into halves, one half using the raw milk and the other half using the pasteurized milk. For those who will be testing the pasteurized milk, carefully pipet exactly 1 ml. of the sample which has been shaken well to break up clumps of organisms, into a 9 ml. sterile water blank and mix the contents well by tapping the base of the tube with your finger. Next, pipet exactly 1 ml. of this 1:10 dilution of the original milk into a second 9 ml. sterile water blank and repeat the mixing. Finally, repeat the procedure one more time with a third sterile water blank to give you a 1:1000 dilution. For those testing the raw milk, perform the same protocol, except carry out the dilutions through a fourth sterile water blank to give a final dilution of 1:10,000 of the original sample.

Each dilution should have a melted and cooled (45°C.) tall of Standard Methods Agar (Tryptone Glucose Yeast Agar) prepared and labeled according to its corresponding dilution. After thoroughly mixing the appropriate dilution, carefully pipet exactly 1 ml. of the dilution into the agar tall and mix well by rotating between the palms of your hands. Pour the contents into a pre-labeled Petri dish and gently swirl the contents to effect an even distribution. After the agar has hardened, incubate inverted in the 37°C. incubator for 48 hours. At that time, count the number of colonies on each plate that shows a range from 30-300 with a colony counter. Record your results for each dilution and calculate the average number of cells per ml. of the original milk sample. The actual dilutions in each Petri plate will be only one-tenth of the dilution from which the sample was pipeted, i.e., pipeting 1 ml. from the 1:10 water balnk would mean a 1:100 dilution in the Petri plate. Thus, if 240 colonies were counted in this plate, this would represent an original bacteria count of 24,000 per ml. of the original sample. Since, due to sampling error, it would not be likely that your counts from each of the dilutions would correspond, average them to get a final determination as to the number of bacteria per ml. of the milk sample and record this value. The average values obtained from other members of the class working with the same sample should likewise be averaged together to get an overall average count. How do the values for pasteurized and raw milk compare?

Since the incubation temperature for the Petri plates in the experiment performed above will only visualize the presence of mesophilic bacteria, a variation on the basic experiment can be performed by setting up the plates described in duplicate. One set could be incubated for a week in the refrigerator and the other in the incubator as prescribed. One could also test for thermophiles by incubating sealed Petri dishes in a 55°C. incubator, but in practice they do not contribute substantially to the total count. The count obtained in the refrigerator would indicate the psychrophilic organisms and their numbers should be added to those obtained in the 37°C. incubator.

As an exercise in some basic microbial ecology, raw milk serves very well to show the effect of environment on the nature of the predominating species. In this case, the sample of milk will be examined after aliquots have been incubated at two different temperature, 20°C. and 37°C. In the former case the temperature will favor the emergence of species of *Streptococcus* and *Lactobacillus* which are responsible for the normal souring (fermentation) of milk, due to the fact that they typically produce lactic acid among the end-products of their fermentation. In the case of the latter temperature, the growth of organisms such as *Escherichia* and *Proteus* is encouraged, giving rise to more complex end-products, many of which will result in foul odors and/or proteolysis (putrefaction) of milk protein.

EXPERIMENT 8-4

The Effect of Temperature on Microbial Growth in Milk

MATERIALS

1. Sample of raw milk.
2. Sterile empty test tubes (2).
3. Endo agar or E.M.B. agar tall (1).
4. Litmus milk (4).

PROCEDURE: Take some of the raw milk sample provided and transfer 5 ml. portions to each of two sterile empty test tubes. Cap the tubes and mark one of them "20°C." and the other "37°C." Transfer each one to the appropriate environment, i.e., the incubator or room temperature. If there is a tendency for the room to warm up much over the 20°C. (68°F.) mark, a suitable incubator should be provided that will maintain a reasonable approximation of the designated temperature.

After 48 hours, remove the tube from the 37°C. incubator. Streak a loopful of the milk (after first swirling well to distribute the organisms in the tube) for isolation on a plate of Endo agar or E.M.B. agar. Note the odor of the tube. Prepare a gram stain on the contents and examine under oil-immersion, making a drawing of the representative types you observe. Finally, inoculate a tube of litmus milk with a loopful of the milk culture and incubate at 37°C. for about a week, making daily observations on the changes taking place.

After 4-5 days, procure the original tube of raw milk incubated at room temperature. Note the consistency of the milk and its odor. Inoculate a tube of litmus milk with the growth from this sample, and incubate it for a week at room temperature, making the usual daily observations on changes that take place. Prepare a gram stain from the raw milk sample just taken from 4-5 days incubation and examine under oil-immersion making appropriate representative drawings of the organisms you see. Are there any obvious predominating types? How do you account for the difference between this gram stain and the one you performed on the 37°C. tube?

CHAPTER 9

SELECTED MEDICAL MICROBIOLOGY LABORATORY PROCEDURES

To manage and treat cases of infectious disease intelligently the physician must know specifically which organism is the causative agent (etiologic agent). In diagnosing infectious diseases, then, it is necessary to isolate and identify the specific organism causing the infection. A great many laboratory procedures have been developed to give this information as accurately and as quickly as possible. In general, clinical laboratory procedures have come about as a result of a very thorough knowledge of the physiological behavior of the different microorganisms causing infectious disease. Tests that identify disease-producing microorganisms demonstrate the value to the medical world of a good knowledge of "microbial physiology."

One of the more commonly employed clinical laboratory tests involves the identification of pathogens associated with diseases of the alimentary canal. In general, this group is collectively called *enteric pathogens*. Some of the diseases caused by organisms of this group include typhoid fever, paratyphoid fever, *Salmonella* food poisoning, and bacterial dysentery. In this country most of the diseases in this group are caused by two genera of bacteria — the *Salmonella* and the *Shigella*. From the study of these organisms some specific test procedures have resulted in which essentially one general procedure is used to detect any of the pathogens found in these two genera.

Many of the most important features in the bacteriological diagnosis of *enteric pathogens* are based on a knowledge of the physiology of these organisms. One of the primary tests in this series relies on the fact that the *Salmonella* and *Shigella* do *not* ferment lactose or sucrose, while most other nonpathogenic, gram-negative inhabitants of the human intestinal tract do ferment either or both of these sugars, producing acid or acid and gas. Therefore, the primary isolation medium should in some way facilitate a differential selection of gram-negative organisms that are nonlactose-fermenting. Various types of media exist that can accomplish this, but you will use just two in the next experiment. One is eosin-methylene-blue agar, or simply E.M.B. agar, and the other is Salmonella-Shigella agar, or S.S. agar. If you will recall, these media are selective for gram-negative bacteria. Both media contain a pH indicator. If a colony of gram-negative bacteria that is capable of fermenting lactose develops on either of these agars, the colony will exhibit color instead of displaying the usual translucence or whitish color seen on normal laboratory media. Sometimes this color envelops the entire colony, and sometimes it simply is concentrated in the center of the colony. Any colony developing on these two media that displays even a central dot of color is considered positive for lactose fermentation. Therefore, the preliminary step in isolation and diagnosis of enteric pathogenic disease involves streaking for isolated colonies with a fecal sample suspected of containing one of these organisms on one or more differential media.

After 24 hours growth, the plates are examined for isolated colonies that have *not* developed color. These colonies presumably may contain the sought-for microorganisms. However, members of the genus *Proteus* may also develop into colonies that appear identical to those of *Salmonella* or *Shigella* since many of these do not ferment lactose. The next step in the procedure involves the inoculation of triple sugar iron (T.S.I.) slants from isolated colonies "picked" from either plate, streaking the surface of the slant and then stabbing the inoculum deep into the medium. Three possible results can be obtained on this medium that indicate a possible pathogen. The culture must display an *acid (yellow) butt* (bottom of the tube) and an *alkaline (red) slant.* The culture is considered *Salmonella*-like (other than *S. typhosa*) if gas bubbles and H_2S (indicated by blackening of the agar) also appear. If H_2S is present (usually only small amounts) but no gas is evident, this is indicative of *Salmonella typhosa*. If neither H_2S nor gas appears, the culture is considered *Shigella*-like. Depending upon the picture obtained here, one proceeds then to eliminate the genus *Proteus*. Organisms of this genus hydrolyze urea; members of *Salmonella* and *Shigella* do not. Urea broth contains phenol red as an indicator. The splitting of urea enzymatically releases carbon dioxide and ammonia, which dissolves in the medium to form ammonium hydroxide. Ammonium hydroxide substantially raises the pH to the alkaline range, and the phenol red shows a full basic color (red-violet or magenta). Nonhydrolyzing bacteria will grow on this medium (turbid appearance) but no change in pH results.

Finally, species identification is carried out. The first step in this procedure is perhaps the most important one. One must obtain a maximum number of well-isolated colonies to be successful in isolating the pathogens, which are not liable to be found in as great numbers as the nonpathogenic organisms. Therefore, if the first attempts result in mostly confluent growth with only a few well-isolated colonies, it is strongly recommended that the student utilize a subdilution technique (prior to streaking) employing sterile water blanks. In this way more well-isolated colonies will be obtained.

Another pitfall is the ever-present problem of a contaminating organism or organisms. Make sure in every step along the way that your preparations are *single, pure cultures* and are not contaminated with others, which may give you incorrect results.

EXPERIMENT 9-1

Isolation and Identification of an Enteric Pathogen

MATERIALS

1. Media (ordinarily the following quantities are adequate for one student):
   a. E.M.B. agar tall (1).
   b. S.S. agar tall (1).
   c. T.S.I. agar slants (4).
   d. Urea broth (4).
   e. Other media necessary for species identification or specific typing serum.
2. Sterile Petri dishes (2).
3. Dilute human fecal suspension containing Salmonella or Shigella species.
4. Specific Salmonella and Shigella typing serum (optional).

PROCEDURE: A dilute fecal sample is provided that has been purposely seeded with either Salmonella or Shigella. These organisms are pathogenic to man, and therefore must be handled with the utmost care. All of the normal precautions you have been taught are sufficient for handling these organisms with safety. It is the person, careless in his technique, who is apt to run into trouble. The procedure for disposing of used media in this exercise is as follows:

For agar plates of all kinds, flood the surface with the disinfectant provided and allow the plates to stand for at least 20 min. before discarding in the usual fashion. For tubed media, either slants or broth cultures of any kind, make sure you bring the medium to a boil in a boiling water bath for at least 5 min. before disposing of it in the usual fashion. All cultures must either be exposed to the disinfectant for at least 20 min. or boiled for 5 min. before disposal.

The sample issued you will be numbered for later identification. You are to report at the conclusion of this exercise the name of the pathogen found in your particular sample. Streak for isolation a plate of E.M.B. agar and a plate of S.S. agar with your sample. After 24 hours, examine both plates for the presence of bacterial colonies that do not show color. These colorless colonies are, presumably, composed of the organism you are looking for. Any colonies that show even so much as a spot of color in the center are usually lactose-fermenting organisms, and therefore nonpathogenic.

With a straight wire inoculating needle, pick two or three well isolated colonies and transfer the growth from each colony to slanted T.S.I. agar. First streak a line of growth upward from the lower portion of the surface of the T.S.I. slant to the upper portion, and then stab to the bottom of the tube (See diagram). Withdraw the needle along the line of each stab. Each colony chosen should be transplanted in this fashion to a separate T.S.I. slant. Mark the colonies picked on the bottom of the Petri plate in some way so that you have a way of knowing which colonies you have already tried. If, after 24 hours growth, the slant shows an acid butt and an alkaline slant with or without the production of H_2S or gas, the presence of either Salmonella or Shigella is indicated. In addition, typical species of the genus Proteus can also give this type of picture. Discard any T.S.I. slant that does not conform to this general pattern.

The next step in the procedure involves the transferring of growth from tubes showing one of the three previously described pictures to a tube of urea broth to rule out the genus Proteus. If, after 24 hours of incubation, the urea broth is not hydrolyzed, you can assume that you are dealing with an enteric pathogen. From this point on, use the identification procedures and charts with which you are familiar. Biochemical identification of various species of Salmonella is very difficult, if not impossible. If you cannot separate your organism from three or four related species, call the organism Salmonella and list all possible species the organism might be. You should be able to identify Salmonella typhosa and most Shigella species. In actual clinical practice, the final identification of a specific organism may be based on a serological identification, i.e., specific typing antiserum, rather than a biochemical one. In your report merely turn in a completed report sheet with your opinion as to which enteric pathogen was present in your sample. Use typing serum if it is available, (spot agglutination test) instead of a biochemical identification.

Another procedure commonly employed in medical laboratories involves the isolation and identification of *pathogenic staphylococci*. It can be assumed that these organisms grow well on ordinary laboratory media, and furthermore that the final definition of a pathogenic staphylococcus involves demonstrating that these organisms are capable of producing the enzyme coagulase. This test is the best criterion for the determination of pathogenicity in these organisms. Most laboratories utilize blood agar as a primary isolation medium, on which observations are made for the presence of colonies of beta-hemolytic staphylococcus without further attempt at a final identification. From this point, suspected colonies can be inoculated into a sample of citrated rabbit or human plasma to determine if they cause clotting or gelling of the plasma. Such a result constitutes a positive coagulase test.

In this experiment you will utilize a medium that is not as frequently employed but is nevertheless quite useful for the isolation of these organisms. This medium is called *mannitol-salt agar*. It is both a differential and a selective medium that can be extremely useful in screening staphylococci from other closely related organisms. It is highly selective because it contains 7.5 per cent sodium chloride, which excludes the growth of most other bacteria. However, micrococci and staphylococci grow very easily on this medium. In addition, this medium contains the fermentable alcohol mannitol, which pathogenic staphylococci generally ferment. Finally, the indicator phenol red is incorporated as a rule, because nonpathogenic staphylococci or micrococci do not ferment mannitol, and therefore colonies of such organisms change phenol red to the full alkaline color (red or purple) or display no color change. Mannitol-fermenting organisms produce a large yellow halo around the colony, indicating fermentation of mannitol. Therefore, the final observation entails looking for large yellow colonies, which are indicative of pathogenic staphylococci. Any red- or purple-colored colonies automatically fall into the presumed category of nonpathogenic micrococci. It has also been noted that pathogenic staphylococci tend to grow quite luxuriantly on this medium, whereas the nonpathogenic organisms appear somewhat inhibited and usually form smaller colonies. One of the excellent features of this medium is that an ordinary cotton swab can be used, rather than an inoculating loop, to streak across the surface of the agar. The fact that virtually none but species of *Micrococcus* or *Staphylococcus* grow on this medium, automatically eliminates other varieties that would undoubtably be on the swab. Final confirmation of pathogenic staphylococci ultimately rests on the demonstration of a positive coagulase test. This will not be done by the student but will be demonstrated using a typical colony from mannitol-salt agar. Various investigators have found that a 96 per cent correlation exists between those colonies appearing to be pathogenic staphylococci on mannitol-salt agar and those producing a positive coagulase test.

EXPERIMENT 9-2

Selective Medium for the Isolation of Pathogenic Staphylococci

MATERIALS

1. Mannitol-salt agar talls (2).
2. Cultures of Staphylococcus aureus, Micrococcus luteus, and Escherichia coli.
3. Citrated plasma for demonstration of coagulase test.
4. Sterile Petri dishes (2).
5. Sterile swab (1).

PROCEDURE: Melt two talls of mannitol-salt agar and pour into Petri plates after they have cooled sufficiently to pour. After the agar has hardened completely, divide each plate into two halves by making a single bisecting mark across the back of the Petri plate. Label each of three halves with the names Staph. aureus, M. luteus, E. coli, and label the fourth one throat swab. Then streak each half plate with the appropriate organism. To streak the throat-swab sector, take a sterile swab and (working with your neighbor) apply this to the back of the throat and then to the appropriate sector. Incubate all plates at 37°C. for 24 to 48 hours. Examine the plates after incubation. Record the results obtained, including: (1) a description of the colony color and size; (2) the color of the zone around the colony; (3) the relative amount of growth, whether it be a small or large amount, using the 0 to 4 system in which the 4 represents maximum growth and 0 no growth; and (4) a description and brief drawing of each of the organisms on a gram stain. Your instructor will perform a coagulase test both on one of the S. aureus colonies that displays a yellow color in the medium and also on a similar appearing colony isolated from a student's throat. What conclusions can you draw regarding the practical use of this mannitol-salt agar medium? Why were the throat swab and E. coli sectors included in this experiment?

The introduction of antigenic materials into the tissues of an animal or human body results in a response on the part of the host. This response takes the form of the production by lymphoid cells of a specific gamma globulin protein, or *antibody*. The formation of antibody globulin is a highly specific process in which the antibody formed will combine only with the antigen that causes its formation, and to a lesser extent, with very close chemical relatives of the original antigenic molecule. You can test for the presence of circulating antibodies in the blood of an experimental animal by several convenient *in vitro* (outside of the body) methods. These methods call for mixing a small sample of serum from the animal and a small amount of the antigen in question and watching for visible signs of combination. The nature of the antigen-antibody combination is such that many times large insoluble complexes of the two components are formed that settle out of the solution (precipitate). It is this "settling out" that can be seen with the naked eye under the appropriate conditions as visible evidence of the reaction between antibody and antigen and is called *agglutination* (if the antigen is cellular) or *precipitation* (if the antigen is molecular).

There are several different ways of showing the agglutination *in vitro* reaction, which in large part is determined by the physical state of the antigen. You will attempt to demonstrate agglutination with one of the more convenient methods available. This test illustrates what happens when specific antibody is mixed with cellular antigens. The formation of large aggregates of cells, clumped or bound together by antibody molecules, is the basis for the agglutination reaction. When this process occurs in the human or animal body, it facilitates the process known as phagocytosis. Various cells in the body are capable of ingesting microorganisms or other foreign debris. It should be obvious that this ingestion process (phagocytosis) is made far more efficient if many cells have been bound together in large clumps prior to incorporation into the phagocyte. Thus, agglutination *in vitro* is representative of a real *in vivo* process. Experimental animals were previously injected with a vaccine consisting of *E. coli* and *Staph. aureus* killed by heat. After a suitable time interval, a sample of blood was removed from the animals and the serum separated from the red and white blood cells. A saline suspension of *E. coli* and *Staph. aureus* was made from an agar slant and this suspension is to be used in this exercise as the antigen. Commercially prepared cell suspensions and antisera can also be used in place of the "homemade" variety.

EXPERIMENT 9-3

The Agglutination Reaction

MATERIALS

1. Serological test tube rack (1).
2. Serological test tubes, 10 mm. x 75 mm. (20).
3. Serum sample from immunized animal, or commercial antiserum.
4. Antigen suspensions* of E. coli and Staph. aureus, or commercial suspension.
5. Saline solution (0.9%).
6. 1.0 ml. pipets (3).

PROCEDURE: Using a wax pencil, number 2 sets of 10 serological tubes each from 1 to 10. Set up tube 10 as an antigen control in each set. Label one set E. coli and the other Staph. aureus. Carefully pipet (aseptic technique is not necessary) 0.8 ml. of saline solution into tube 1 in each set, and 0.5 ml. of saline solution into each of the other tubes. Pipet 0.2 ml. of the undiluted serum into tube 1 of each set. Using a separate pipet for each set of the two sets of tubes, begin your serial dilution of the serum. This is accomplished by alternately sucking up most of the contents of the tube into your pipet and gently blowing it back into the tube five times for each tube. This is to insure a good mixture of the serum and the saline. After tube 1

---

*Note: A usable antigen suspension of bacteria may be prepared by adding 3 to 4 ml. sterile 0.9% saline to each nutrient agar slant containing a 24-hour culture of the organism. Gently shake the contents to wash off all growth, then pour into a sterile container. Heat the pooled suspension of several nutrient agar slants in a water bath set at 65°C. for 1 hour to kill the organisms. For immunizing a rabbit, this same suspension may be used, providing the killed cells have been washed several times by centrifuging in 0.9% saline to remove any soluble toxins. The following schedule is usually adequate:

        Day 1 = 0.1 ml. (intravenous injection, marginal ear vein)
        Day 3 = 0.2 ml. (intravenous injection, marginal ear vein)
        Day 5 = 0.3 ml. (intravenous injection, marginal ear vein)
        Day 7 = 0.5 ml. (intravenous injection, marginal ear vein)
        Day 14 = Bleed.

This may be accomplished by cardiac puncture or slicing into the marginal ear vein of an anesthetized rabbit with a razor blade and collecting the blood in an open flask. If the latter method is used, pre-treat the area with some toluene to retard clotting.

Refrigerate the blood for 2-3 hours, then aspirate serum, or centrifuge first at about 2000 r.p.m. for 10 minutes to pack the clot; then remove serum.

has been well mixed, transfer 0.5 ml. of the mixture into the next tube (tube 2) and repeat the mixing process. Repeat this process in all 9 tubes — that is, mixing the contents and then pipeting 0.5 ml. of this mixture into the next tube of the series. Discard 0.5 ml. from the last tube after it has been mixed well. Don't add serum to the antigen control tubes (tube 10). An antigen control must be set up for each of the series to make sure that the antigen suspension does not agglutinate spontaneously and also to have a nonagglutinating standard with which to compare those tubes that display visible agglutination. Into each control tube, pipet 0.5 ml. of saline solution followed by 0.5 ml. of the antigen suspension. To each of the serum dilutions, add 0.5 ml. of the antigen suspension that corresponds to your labeling, i.e., E. coli suspension added to one of the series and the Staph. aureus suspension added to the other. The contents of each tube and the resulting final dilutions are shown in the chart below:

Tube #	Serum	Saline	Antigen	Dilution
1	0.2 ml.	0.8 ml.	0.5 ml.	1:10
2		0.5 ml.	0.5 ml.	1:20
3		0.5 ml.	0.5 ml.	1:40
4		0.5 ml.	0.5 ml.	1:80
5		0.5 ml.	0.5 ml.	1:160
6		0.5 ml.	0.5 ml.	1:320
7		0.5 ml.	0.5 ml.	1:640
8		0.5 ml.	0.5 ml.	1:1280
9		0.5 ml.	0.5 ml.	1:2560
10	None	0.5 ml.	0.5 ml.	Antigen Control

Mix all tubes well by shaking the tube rack gently, and then immerse the tubes and rack in a constant temperature water bath or incubate at 37°C. for 2 to 4 hours. Remove the rack from the water bath and put it into a refrigerator overnight. (You may centrifuge the tubes for 20 min. at 2500 r.p.m. after the water bath instead of using overnight refrigeration, if desired). In reading the results, start with the antigen control tube. Gently tap the base of this tube several times with finger tip, and observe the manner in which these cells become resuspended in the solution. This tube should show immediate dispersal of the bacteria in the saline solution with no evident clumping. Repeat the procedure with tube 1 and observe the large clumps in contrast to the control tube. Continue checking each of the tubes until you determine the last tube that shows any visible clumping. The highest (greatest) dilution at which clumping of the cells can be observed is recorded as the <u>titer</u> of the antiserum. For example, if tube 4 is the last one to show clumping, the titer of the serum would be reported as 1:80. Tabulate your results for each of the tubes in the experiment, recording good agglutination as 2+, questionable agglutination as ± and no agglutination as −. Record the titer of the antiserum.

EXPERIMENT 9-4

The Morphology of Blood Cells, Erythrocyte Blood Typing and Differential Count

MATERIALS

1. Clean microscope slides (4).
2. Skin antiseptic.
3. Sterile blood lancet (1).
4. ABO typing antiserum.
5. Wright's blood stain.
6. Wright's buffer.
7. Test tubes 13 × 100 mm., containing 0.5 ml. sterile 0.85% saline (1).

PROCEDURE: Prepare two clean slides, one for blood typing and the other for a blood cell morphology and differential white cell count. On the underside of the blood typing slide, mark an "A" on one area and a "B" on another with your glass marking pencil. Also

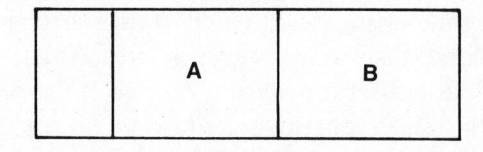

A	B

mark vertical lines as indicated, to delimit the appropriate area. Place a second clean slide next to the one above. Now swab the palmar surface of your fingers with a suitable antiseptic and allow it to soak for at least 1 minute or more (remember, bacteria are not killed instantly!). Next, make a quick stab into the cleaned skin with a sterile blood lancet about 1-2 mm. deep (it is often easier to let your lab partner do it for you). Express a drop of blood on each of the A and B areas of the first slide, followed by a drop near end of the

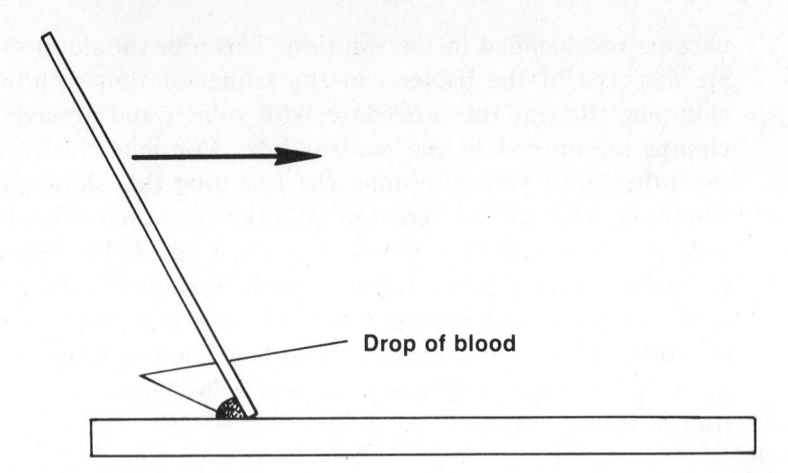

Figure 9-1 The tilted slide is "backed" into the drop of blood until the fluid spreads the width of the slide. Then the blood is dragged across the length of the bottom slide with a slow steady motion.

second slide. Finally, put the finger with the incision over the end of the tube with the saline solution, then set it aside. Quickly apply one drop of anti-A antiserum to the drop of blood in area A of the first slide, making sure that no blood touches the serum from the dropper. Repeat with anti-B antiserum on area B. Next, take a clean slide and, placing it at a 45-degree angle, back into the drop of blood at one end of the second slide, then push it across the length of the slide after the blood has spread to the width of the glass by capillarity (see Figure 9-1).

The resulting thin smear of blood should be allowed to air-dry. While this is taking place, begin rocking your A and B slide back and forth along its length to mix the blood and antiserum which will appear as large, "grainy" red clumps that get progressively larger in a positive reaction. A non-agglutination reaction will remain a homogeneous red-colored suspension. Record blood type. Next, prepare another slide with areas A and B as above, putting a drop of the saline blood suspension on each area, followed by a drop of the appropriate antiserum (anti-A in area A, anti-B in area B). Mix by tilting the slide lengthwise, place a cover-slip on the drop, then examine each area with the low power objective of your microscope, observing for agglutination of individual cells. Make a drawing of representative cells from each area. (If you are type-O, borrow a slide from someone with agglutination for one of your drawings and indicate its type.)

The blood smear you prepared earlier will be stained to permit a visualization and differentiation of the various blood cell types. Fill a dropper with Wright's blood stain and apply exactly 15 drops to the smear rapidly, spreading the stain over the smear with the dropper.

After 2 minutes have passed (time this procedure carefully), add exactly 15 drops of Wright's buffer or distilled water to the stain and mix the stain and buffer by tilting the slide lengthwise. Do not pour off the Wright's stain before adding buffer or water — mix the two together. After 2 more minutes have elapsed, rinse the slide with tap water, blot dry and examine under the oil-immersion lens. Make a drawing of each representative cell type (erythrocyte, monocyte, lymphocyte, neutrophil, eosinophil and basophil).

Finally, perform a differential white blood cell count on the stained slide. For this procedure, begin near one end of your smear and travel straight across to the other end of the slide, tallying each white blood cell you encounter until you have counted 200 total cells:

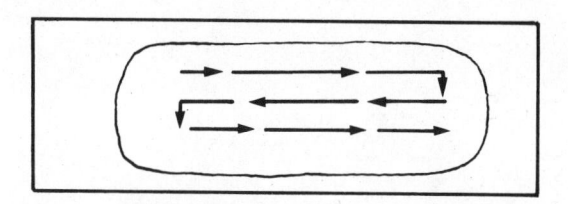

When you have completed this, calculate the percentage representation of each cell type and record this figure. For example, if you counted 40 lymphocytes among the 200, this would represent 20 per cent of the total WBC population: 40/200 = .20. How do your figures compare with accepted normal values? How could you improve the accuracy of your figures?

NAME _____

DATE _____

SECTION _____

EXPERIMENT 1-1

Introduction to the Use of the Microscope

PURPOSE: _____

_____

DATA:

1. Human hair.
2. Wet mount.
3. 100 X.

1. Human hair.
2. Wet mount.
3. 450 X.

1. Cotton fibers.
2. Wet mount.
3. 100 X.

1. Cotton fibers.
2. Wet mount.
3. 450 X.

1. Torn edge of paper.
2. Dry mount.
3. 100 X.

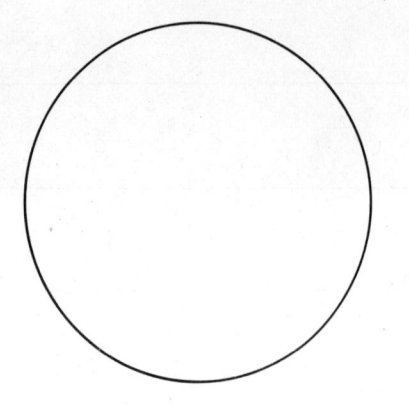

1. Torn edge of paper.
2. Dry mount.
3. 450 X.

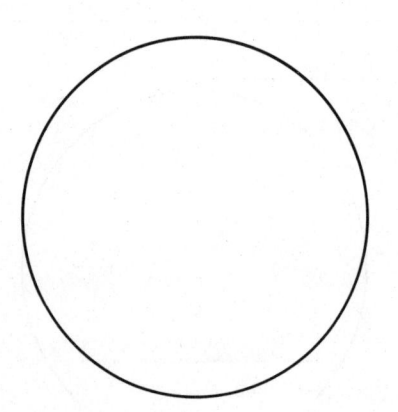

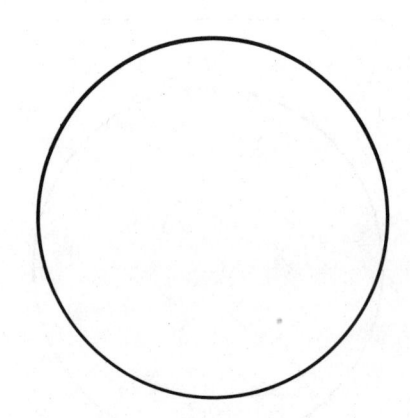

1. NaCl (salt) crystals.
2. Dry preparation.
3. 100 X.

1. NaCl (salt) crystals.
2. Dry preparation.
3. 450 X.

DISCUSSION AND CONCLUSIONS: _____

NAME _____

DATE _____

SECTION _____

EXPERIMENT 1-2

The Use of Pipets, Indicators and Buffers

PURPOSE: _____

_____

DATA:

Tube	Tube Contents	No. of Drops	Observations
2	K_2HPO_4 buffer		
3	KH_2PO_4 buffer		
4	8% Peptone solution		

DISCUSSION AND CONCLUSIONS: _____

_____

_____

_____

_____

_____

_____

NAME _____

DATE _____

SECTION _____

EXPERIMENT 1-4

The Omnipresence of Microorganisms in the Environment and the Necessity for Aseptic (Sterile) Technique

PURPOSE: _____

_____

_____

DATA:

A. Plate 1: nonsterile plate poured with agar. _____

_____

_____

B. Plate 2: sterile plate with agar exposed to the air. _____

_____

_____

C. Test tubes, with broth transferred without using aseptic technique.

Tube 1: _____

Tube 2: _____

D. Colony study

	Colony (sketch)	Form	Elevation	Margin
1				
2				
3				

DISCUSSION AND CONCLUSIONS: _____

_____

_____

_____

_____

_____

_____

_____

_____

_____

_____

NAME _____

DATE _____

SECTION _____

EXPERIMENT 1-5

The Cultivation of Anaerobic Organisms

PURPOSE: _____
_____

DATA:

Organism	Medium	Results (growth + or −)	Growth Pattern (sketch)
Clostridium sporogenes	Thioglycollate broth		
	Nutrient agar tall		
	Nutrient agar slant		
Bacillus subtilis	Thioglycollate broth		
	Nutrient agar tall		
	Nutrient agar tall		

DISCUSSION AND CONCLUSIONS:

NAME _____

DATE _____

SECTION _____

EXPERIMENT 2-1

Smear Preparations and Fixation

PURPOSE: _____

_____

_____

DATA:

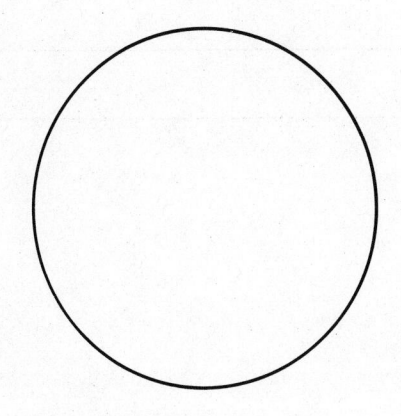

1. Microbial culture.
2. Loeffler's methylene blue (fixed with heat).
3. Oil-immersion.

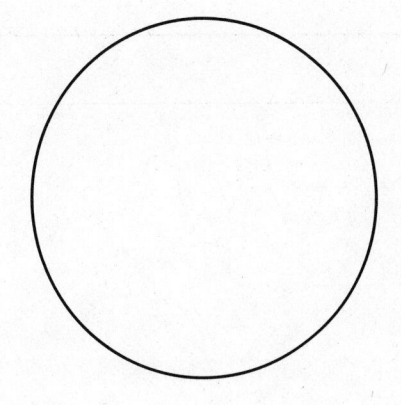

1. Imbedded material from between teeth.
2. Loeffler's methylene blue (fixed with heat).
3. Oil-immersion.

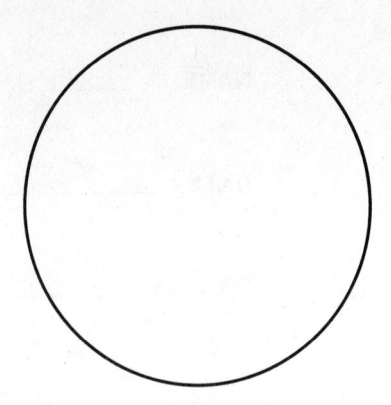

○

1. Scrapings from inner lining of cheek.
2. Loeffler's methylene blue (fixed with alcohol).
3. Oil-immersion.

DISCUSSION AND CONCLUSIONS: _____

_____

_____

_____

_____

_____

_____

_____

NAME _____

DATE _____

SECTION _____

EXPERIMENT 2-2

Negative Staining of Microorganisms

PURPOSE: _____
_____
_____

DATA:

1. Mixed suspension.
2. Dorner's nigrosin (not fixed).
3. Oil-immersion.

1. Mixed suspension.
2. Congo Red (not fixed).
3. Oil-immersion.

161

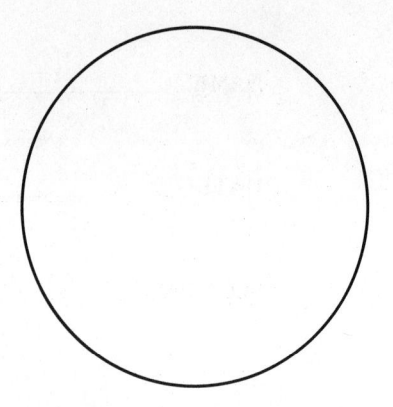

○

1. Mixed suspension.
2. Methylene blue (heat-fixed).
3. Oil-immersion.

DISCUSSION AND CONCLUSIONS: _____

_____
_____
_____
_____
_____
_____
_____

NAME _____

DATE _____

SECTION _____

EXPERIMENT 2-3

Motility Studies on Microorganisms

PURPOSE: _____

_____

DATA:

        Hanging drop preparations of each of the following:

Slide 1 <u>Micrococcus luteus.</u> _____

_____

Slide 2 <u>Bacillus subtilis.</u> _____

_____

Slide 3 <u>Hay infusion.</u> _____

_____

DISCUSSION AND CONCLUSIONS: _____

_____

_____

_____

_____

_____

_____

NAME _____

DATE _____

SECTION _____

EXPERIMENT 2-4

The Demonstration of Fat Inclusion Bodies

PURPOSE: _____
_____
_____

DATA:

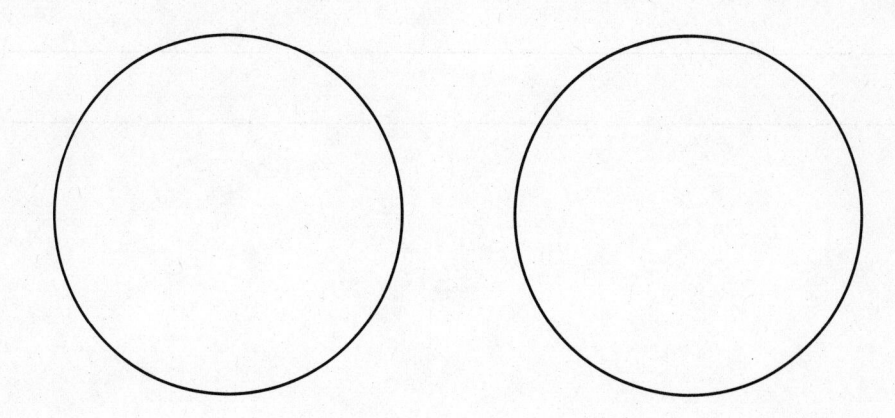

1. <u>Bacillus subtilis.</u>
2. Sudan black B and safranin.
3. Oil-immersion.

1. <u>Saccharomyces cerevisiae.</u>
2. Sudan black B (wet mount).
3. Oil-immersion.

165

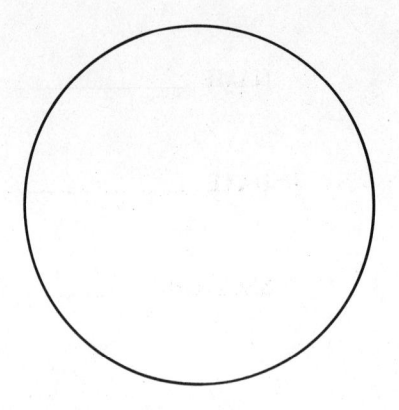

1. <u>Saccharomyces cerevisiae.</u>
2. Lugol's iodine solution (wet mount).
3. Oil-immersion.

DISCUSSION AND CONCLUSIONS: _____

_____

_____

_____

_____

_____

_____

NAME _____

DATE _____

SECTION _____

EXPERIMENT 2-5

The Demonstration of Metachromatic Granules

PURPOSE: _____

_____

DATA:

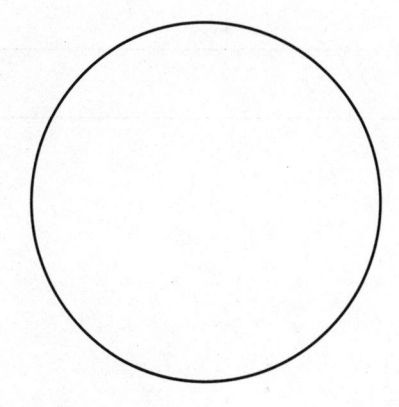

○   ○

1. <u>Corynebacterium
   pseudodiphtheriticum.</u>
2. Albert's diphtheria stain.
3. Oil-immersion.

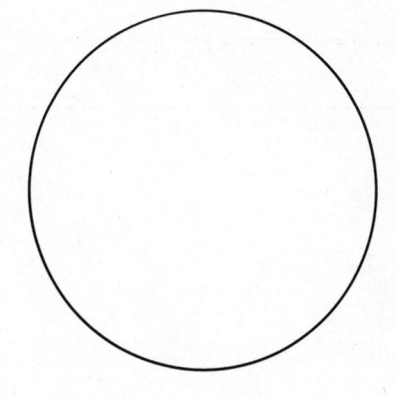

1. <u>Corynebacterium
   pseudodiphtheriticum.</u>
2. Loeffler's methylene blue.
3. Oil-immersion.

○

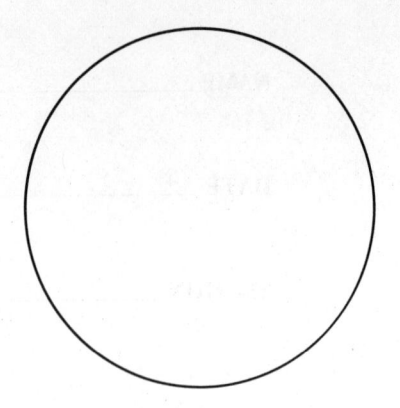

1. Hay infusion.
2. Albert's stain.
3. Oil-immersion.

DISCUSSION AND CONCLUSIONS: _____

_____

_____

_____

_____

_____

_____

NAME _____

DATE _____

SECTION _____

EXPERIMENT 2-6

The Gram Stain

PURPOSE: _____

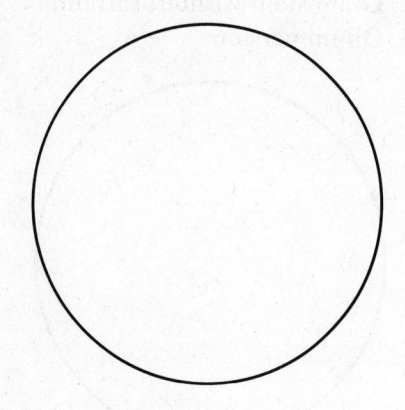

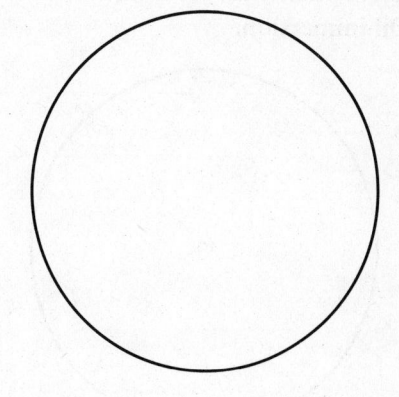

_____

_____

DATA:

Slide 1

1. Smear from gums.
2. Complete gram stain.
3. Oil-immersion.

1. Staphylococcus epidermidis
2. Complete gram stain.
3. Oil-immersion.

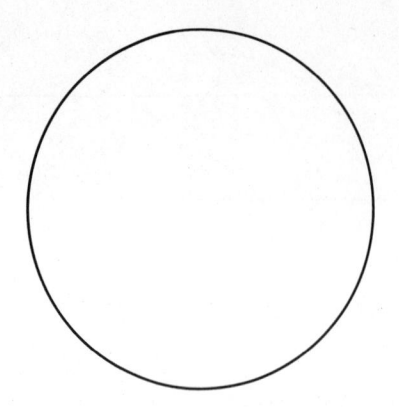

1. Saccharomyces cerevisiae.
2. Complete gram stain.
3. Oil-immersion.

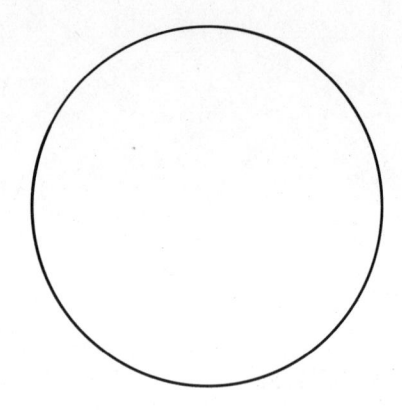

1. Escherichia coli.
2. Complete gram stain.
3. Oil-immersion.

Slide 2

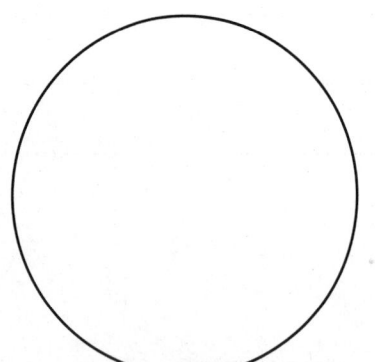

1. Smear from gums.
2. Gram stain without safranin.
3. Oil-immersion.

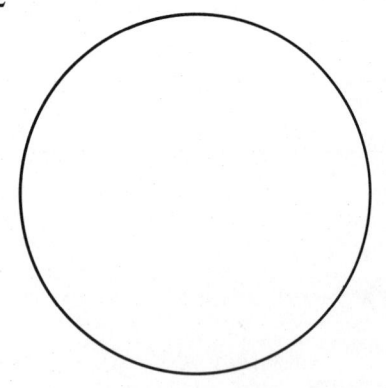

1. Staphylococcus epidermidis.
2. Gram stain without safranin.
3. Oil-immersion.

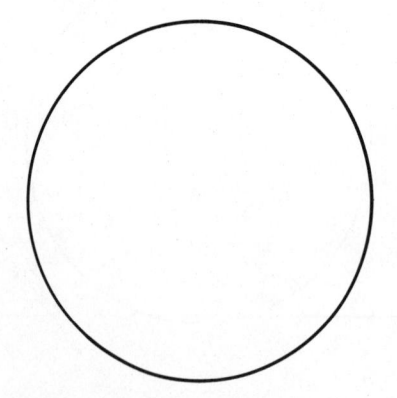

1. Saccharomyces cerevisiae.
2. Gram stain without safranin.
3. Oil immersion.

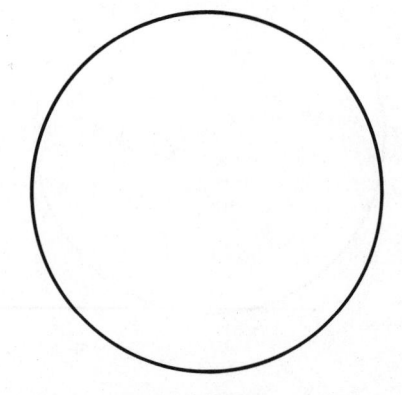

1. Escherichia coli.
2. Gram stain without safranin.
3. Oil-immersion.

Slide 3

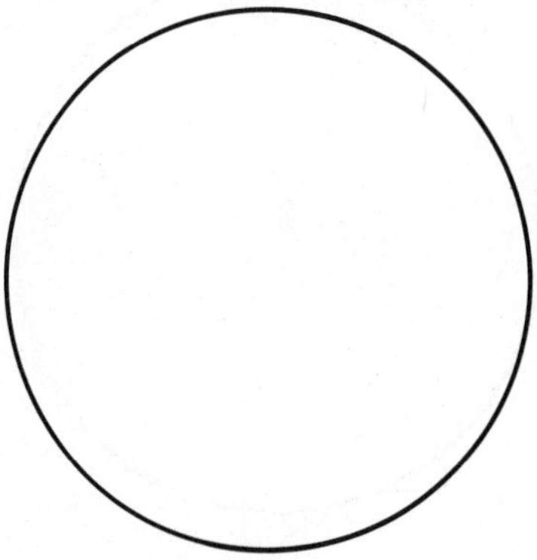

1. Smear from gums.
2. Gram's crystal violet and Lugol's solution.
3. Oil-immersion.

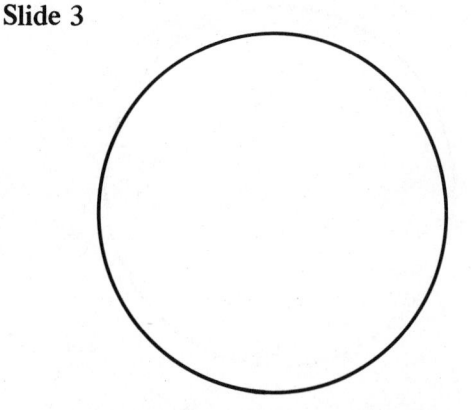

1. Staphylococcus epidermidis.
2. Gram's crystal violet and Lugol's solution.
3. Oil-immersion.

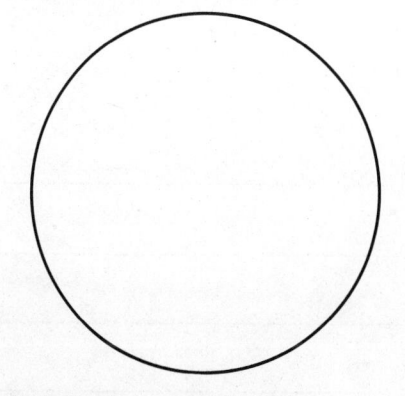

1. Saccharomyces cerevisiae.
2. Gram's crystal violet and Lugol's solution.
3. Oil-immersion.

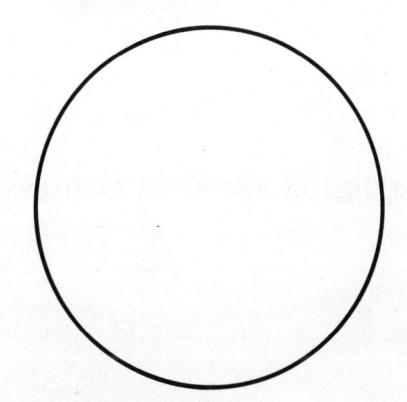

1. Escherichia coli.
2. Gram's crystal violet and Lugol's solution.
3. Oil-immersion.

Slide 4

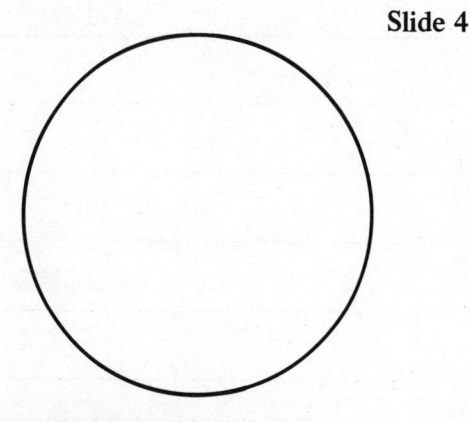

1. Smear from gums.
2. Gram's crystal violet only.
3. Oil-immersion.

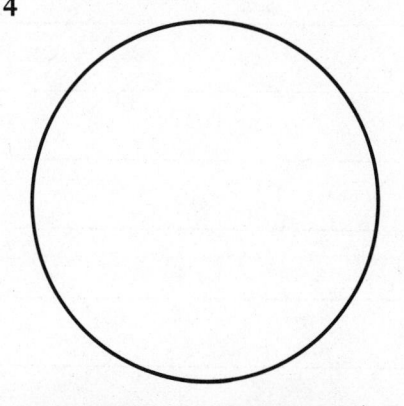

1. Staphylococcus epidermidis.
2. Gram's crystal violet only.
3. Oil-immersion.

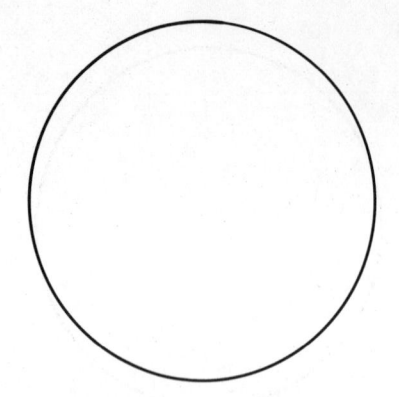

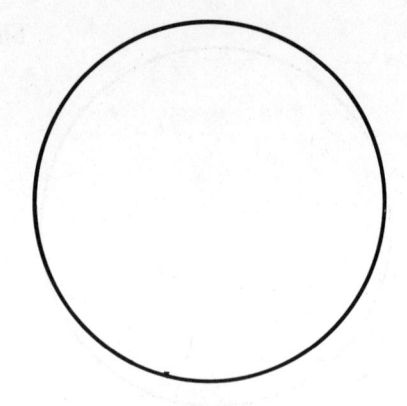

1. Saccharomyces cerevisiae.
2. Gram's crystal violet only.
3. Oil-immersion.

1. Escherichia coli.
2. Gram's crystal violet only.
3. Oil-immersion.

DISCUSSION AND CONCLUSIONS: _____

NAME _____

DATE _____

SECTION _____

EXPERIMENT 2-7

The Acid-fast Stain

PURPOSE: _____

_____

DATA:

○   ○

1. <u>Mycobacterium tuberculosis</u> and <u>Staphylococcus epidermidis.</u>
2. Acid-fast stain.
3. Oil-immersion.

1. Tuberculous sputum.
2. Acid-fast stain.
3. Oil-immersion.

DISCUSSION AND CONCLUSIONS: _____

_____
_____
_____
_____
_____

NAME _____

DATE _____

SECTION _____

EXPERIMENT 2-8

Demonstration of Bacterial Endospores

PURPOSE: _____

_____

DATA:

1. <u>Bacillus species.</u>
2. Loeffler's methylene blue.
3. Oil-immersion.

1. <u>Bacillus species.</u>
2. Schaeffer-Fulton method.
3. Oil-immersion.

DISCUSSION AND CONCLUSIONS: _____

_____

_____

_____

_____

_____

_____

175

NAME _____

DATE _____

SECTION _____

EXPERIMENT 2-11

Demonstration of the Bacterial Cell Wall

PURPOSE: _____
_____
_____

DATA:

      1. Bacillus subtilis.
      2. Phosphomolybdic acid (1%) and methyl green (1%).
      3. Oil-immersion.

DISCUSSION AND CONCLUSIONS: _____
_____
_____
_____
_____
_____

NAME _____

DATE _____

SECTION _____

EXPERIMENT 2-12

Demonstration of the Nucleus and General Morphology of Yeasts

PURPOSE: _____
_____

DATA:

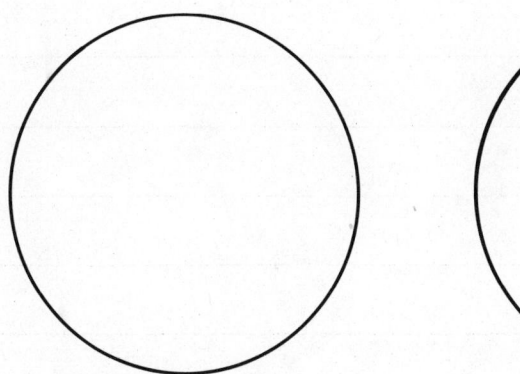

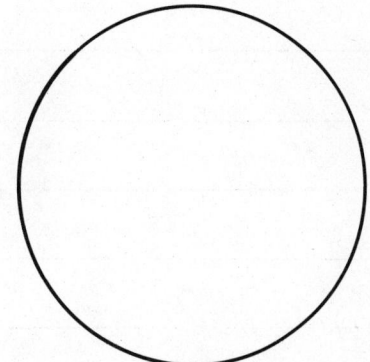

1. Saccharomyces cerevisiae.	1. Saccharomyces cerevisiae.
2. Toluidine blue (0.1%) (with potassium hydroxide).	2. Toluidine blue (0.1%) (without potassium hydroxide).
3. Oil-immersion.	3. Oil-immersion.

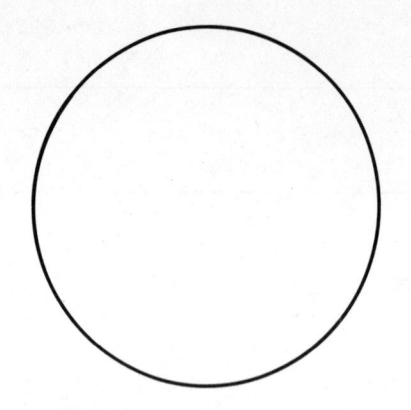

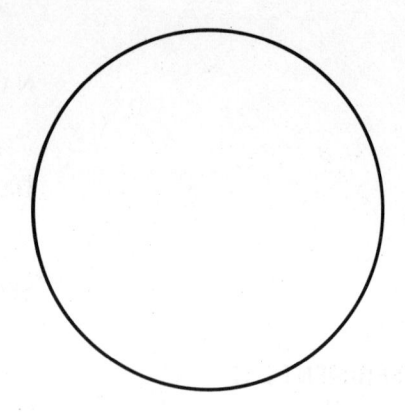

1. <u>Saccharomyces cerevisiae.</u>
2. Gram's iodine (wet mount).
3. High power.

1. <u>Saccharomyces cerevisiae.</u>
2. Gram's iodine (wet mount).
3. Oil-immersion.

DISCUSSION AND CONCLUSIONS: _____

_____
_____
_____
_____
_____
_____
_____
_____
_____

NAME _____

DATE _____

SECTION _____

EXPERIMENT 2-13

A Study of Mold Morphology

PURPOSE: _____
_____

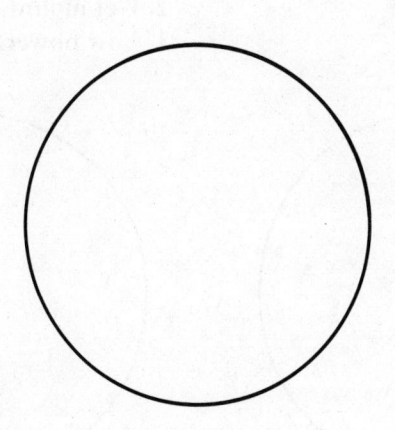

DATA:

○

1. <u>Aspergillus niger.</u>
2. Microculture.
3. Low power.

○

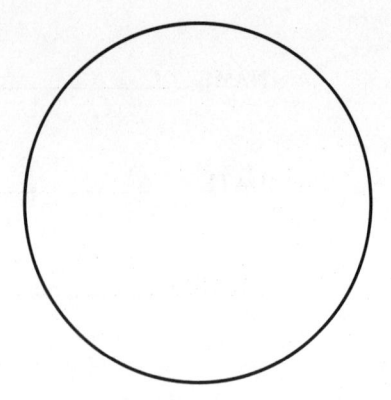

1. <u>Penicillium notatum.</u>
2. Microculture.
3. Low power.

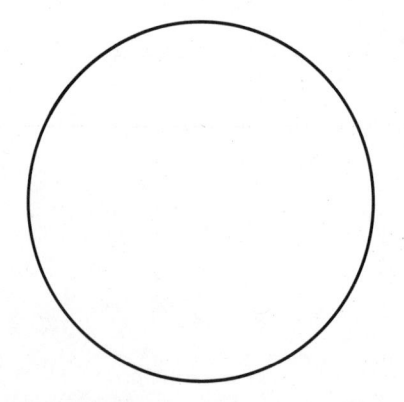

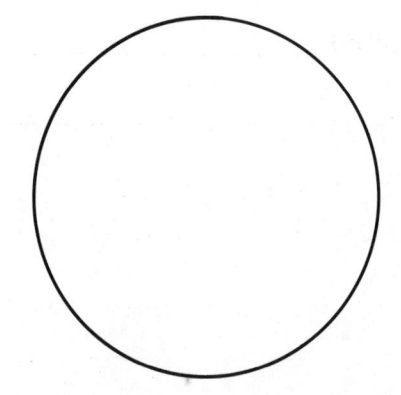

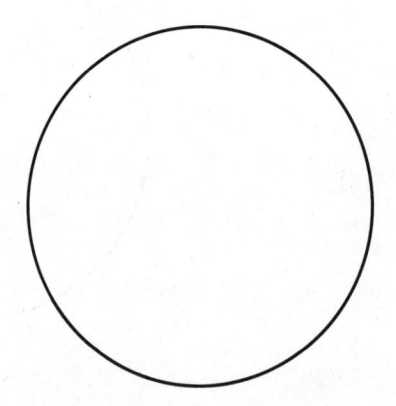

○ ○

1. Mold from air (color of spores, if any).
2. Wet mount.
3. Low power.

1. Mold from air (color of spores, if any).
2. Wet mount.
3. Low power.

○ ○

1. Mold from air (color of spores, if any).
2. Wet mount.
3. High power.

1. Mold from air (color of spores, if any).
2. Wet mount.
3. High power.

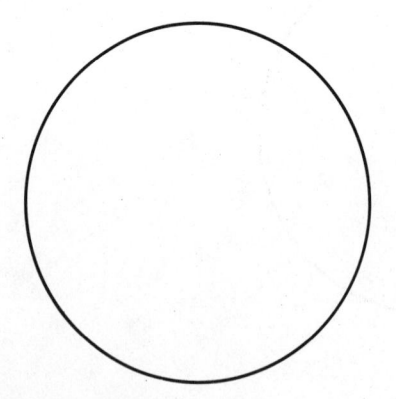

DISCUSSION AND CONCLUSIONS: _____

NAME _____

DATE _____

SECTION _____

EXPERIMENT 2-14

The Study of Protozoan Morphology

PURPOSE: _____
_____

DATA:

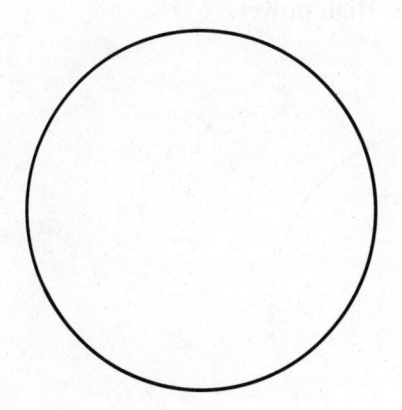

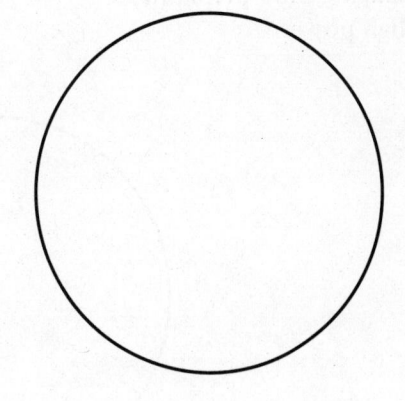

1. <u>Amoeba proteus.</u>	1. <u>Paramecium caudatum.</u>
2. Prepared slide.	2. Prepared slide.
3.	3.

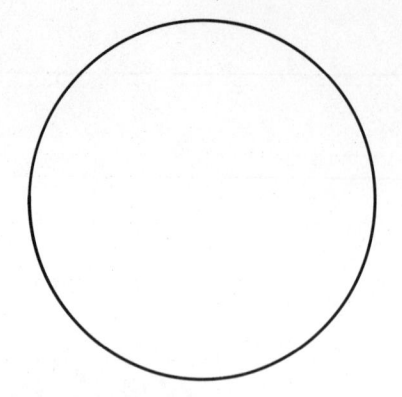

1. Euglena viridis.
2. Prepared slide.
3.

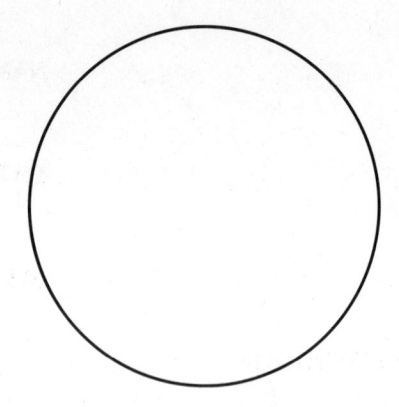

1. Plasmodium malariae.
2. Prepared slide.
3.

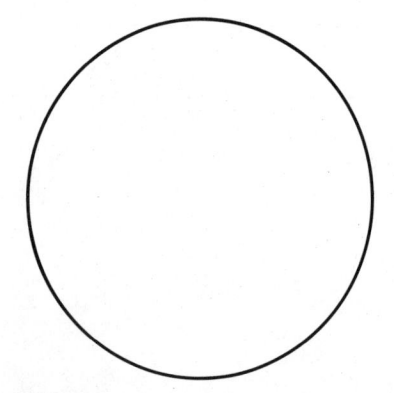

1. Two different species from hay infusion.
2. Hanging drop preparation.
3. High power.

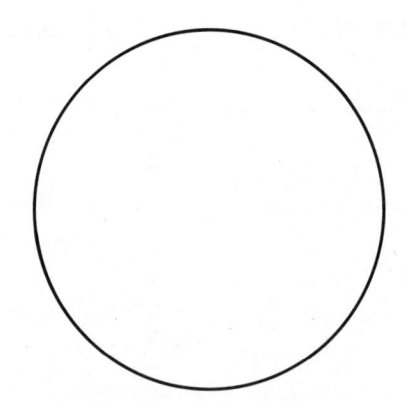

1. Paramecium in fission.
2. Prepared slide.
3. High power.

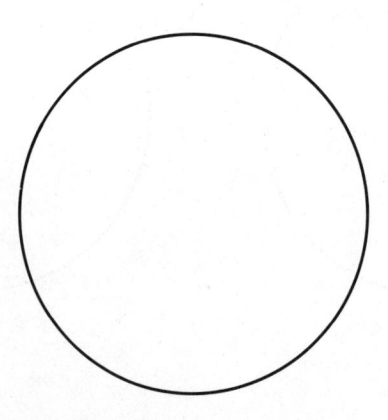

1. Paramecium in conjugation.
2. Prepared slide.
3. High power.

DISCUSSION AND CONCLUSIONS:

NAME _____

DATE _____

SECTION _____

EXPERIMENT 2-15

Cytological Characterization of an Unknown Bacterium

PURPOSE: _____

_____

DATA:

Number of the Unknown Bacterium: _____.

    Results from the following stains or procedures:

1. Fat inclusion bodies. _____

2. Metachromatic granules. _____

3. Gram stain. _____

4. Acid-fast stain. _____

5. Spore stain. _____

6. Flagella stain. _____

7. Capsule demonstration. _____

8. Motility preparation. _____

NAME _____

DATE _____

SECTION _____

EXPERIMENT 3-1

Determination of Minimal Growth Requirements

PURPOSE: _____
_____

DATA:

Plate 1. Agar only.

    Sector 1 <u>Escherichia coli</u> _____

    Sector 2 <u>Streptococcus lactis</u> _____

    Sector 3 <u>Saccharomyces cerevisiae</u> _____

    Sector 4 Control _____

Plate 2. Agar + minerals.

    Sector 1 <u>Escherichia coli</u> _____

    Sector 2 <u>Streptococcus lactis</u> _____

    Sector 3 <u>Saccharomyces cerevisiae</u> _____

    Sector 4 Control _____

Plate 3. Agar + minerals + organic carbon source.

    Sector 1 <u>Escherichia coli</u> _____

    Sector 2 <u>Streptococcus lactis</u> _____

    Sector 3 <u>Saccharomyces cerevisiae</u> _____

    Sector 4 Control _____

Plate 4. Agar + minerals + organic carbon source + organic nitrogen source.

    Sector 1 <u>Escherichia coli</u> _____

    Sector 2 <u>Streptococcus lactis</u> _____

    Sector 3 <u>Saccharomyces cerevisiae</u> _____

    Sector 4 Control _____

DISCUSSION AND CONCLUSIONS: _____

_____

_____

_____

_____

_____

_____

_____

_____

_____

_____

NAME _____

DATE _____

SECTION _____

EXPERIMENT 3-2

Selective and Differential Media

PURPOSE: _____

_____

_____

DATA:

Plate 1. Nutrient agar.

    Sector 1 Staphylococcus aureus   _____

    Sector 2 Alcaligenes faecalis   _____

    Sector 3 Control   _____

Plate 2. Sodium chloride agar.

    Sector 1 Staphylococcus aureus.   _____

    Sector 2 Alcaligenes faecalis   _____

    Sector 3 Control   _____

Plate 3. Phenol red agar.

    Sector 1 Staphylococcus aureus   _____

    Sector 2 Alcaligenes faecalis   _____

    Sector 3 Control   _____

DISCUSSION AND CONCLUSIONS:

NAME _____

DATE _____

SECTION _____

EXPERIMENT 3-3

Utilization of Unusual Sources of Nitrogen and Carbon by Microorganisms

PURPOSE: _____

_____

_____

DATA:

Tube 1. Escherichia coli in ammonium phosphate broth.

_____

Tube 2. Aerobacter aerogenes in ammonium phosphate broth.

_____

Tube 3. Escherichia coli in sodium citrate broth (or Simmon's citrate agar slant).

_____

Tube 4. Aerobacter aerogenes in sodium citrate broth (or Simmon's citrate agar slant).

_____

Tube 5. Escherichia coli in uric acid broth.

_____

Tube 6. Aerobacter aerogenes in uric acid broth.

_____

DISCUSSION AND CONCLUSIONS:

NAME _____

DATE _____

SECTION _____

EXPERIMENT 4-1

The Hydrolysis of Starch

PURPOSE: _____
_____

DATA:

1. Petri dish with starch agar flooded with iodine solution.

	Escherichia coli	Bacillus subtilis
Starch Hydrolysis		

DISCUSSION AND CONCLUSIONS: _____
_____
_____
_____
_____

NAME _____

DATE _____

SECTION _____

EXPERIMENT 4-2

Hydrolysis of Casein

PURPOSE: _____
_____

DATA:

Petri dish with milk agar.

	Escherichia coli	Bacillus subtilis
Casein Hydrolysis		

DISCUSSION AND CONCLUSIONS: _____
_____
_____
_____
_____

203

NAME _____

DATE _____

SECTION _____

EXPERIMENT 4-3

The Hydrolysis of Gelatin

PURPOSE: _____
_____

DATA:

Plate 1. Escherichia coli on 0.4% nutrient gelatin agar. _____
_____

Plate 2. Bacillus subtilis on 0.4% nutrient gelatin agar. _____
_____

Plate 3. Proteus vulgaris on 0.4% nutrient gelatin agar. _____
_____

Tube 1. Escherichia coli in stab culture of nutrient gelatin. _____
_____

Tube 2. Bacillus subtilis in stab culture of nutrient gelatin. _____
_____

Tube 3. Proteus vulgaris in stab culture of nutrient gelatin. _____
_____

DISCUSSION AND CONCLUSIONS:

NAME _____

DATE _____

SECTION _____

EXPERIMENT 4-4

Hemolysin Production

PURPOSE: _____

_____

DATA:

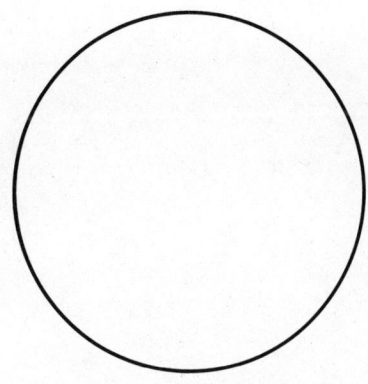

Petri dish with blood agar.

Organism	Staph. aureus or B. subtilis	D. pneumoniae or Strep. lactis
Type of Hemolysis		

207

DISCUSSION AND CONCLUSIONS: _____

_____
_____
_____
_____
_____

NAME _____

DATE _____

SECTION _____

EXPERIMENT 4-5

Coagulase Production (Demonstration)

PURPOSE: _____
_____
_____

DATA:

Tube 1. Citrated plasma inoculated with Staphylococcus aureus after incubation.

_____

Tube 2. Citrated plasma not inoculated (used as a control) after incubation.

_____

DISCUSSION AND CONCLUSIONS: _____
_____
_____
_____
_____
_____

NAME _____

DATE _____

SECTION _____

EXPERIMENT 4-6

Alcohol Production by Yeasts

PURPOSE: _____

_____

_____

DATA:

Tube 1. Saccharomyces cerevisiae in malt extract broth.

_____

Tube 2. Saccharomyces cerevisiae in distilled water with mashed raisins.

_____

Tube 3. Uninoculated and unheated mashed raisins in distilled water.

_____

DISCUSSION AND CONCLUSIONS: _____

_____

_____

_____

_____

NAME _____

DATE _____

SECTION _____

EXPERIMENT 4-7

Lactic Acid Production by Bacteria

PURPOSE: _____
_____
_____

DATA:

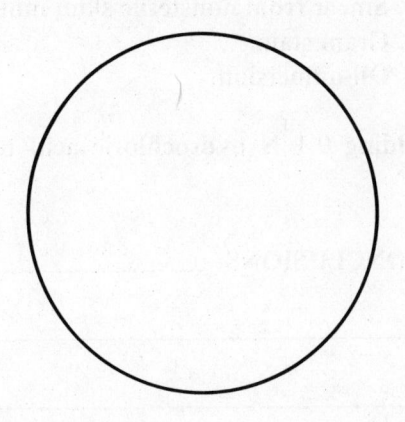

1. Smear from sterile skim milk inoculated with <u>Lactobacillus casei.</u>
2. Gram stain.
3. Oil-immersion.

213

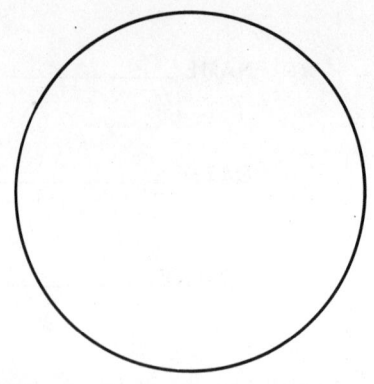

◯

1. Smear from sterile skim milk inoculated with <u>Streptococcus lactis.</u>
2. Gram stain.
3. Oil-immersion.

◯

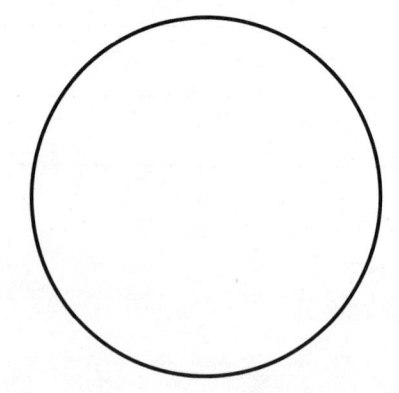

1. Smear from nonsterile skim milk.
2. Gram stain.
3. Oil-immersion

Observations after adding 0.1 N hydrochloric acid to the nonsterile skim milk.

DISCUSSION AND CONCLUSIONS: _____

_____
_____
_____
_____
_____

NAME _____

DATE _____

SECTION _____

EXPERIMENT 4-8

Fermentation of Carbohydrates by Microorganisms

PURPOSE: _____

_____

DATA

Microorganism	Broth	24 Hours	48 Hours
Escherichia coli	Glucose		
	Lactose		
	Sucrose		
Staphylococcus aureus	Glucose		
	Lactose		
	Sucrose		
Alcaligenes faecalis	Glucose		
	Lactose		
	Sucrose		

DISCUSSION AND CONCLUSIONS: _____

_____

_____

_____

NAME _____

DATE _____

SECTION _____

EXPERIMENT 4-9

Reduction of Methylene Blue

PURPOSE: _____
_____

DATA:

Tube 1. 6 ml. of broth culture and 3 ml. of sterile nutrient broth.
_____

Tube 2. 4.5 ml. of broth culture and 4.5 ml. of sterile nutrient broth.
_____

Tube 3. 3 ml. of broth culture and 6 ml. of sterile nutrient broth.
_____

DISCUSSION AND CONCLUSIONS: _____
_____
_____
_____
_____
_____
_____

NAME _____

DATE _____

SECTION _____

EXPERIMENT 4-10

Reduction of Litmus in Milk

PURPOSE: _____
_____
_____

DATA

Tube 1: litmus milk inoculated with Escherichia coli.

Day	Litmus Reduction	Curd	Protein Hydrolysis	Gas Production	pH Change

Tube 2: litmus milk inoculated with Bacillus subtilis.

Day	Litmus Reduction	Curd	Protein Hydrolysis	Gas Production	pH Change

Tube 3: litmus milk inoculated with Streptococcus lactis.

Day	Litmus Reduction	Curd	Protein Hydrolysis	Gas Production	pH Change

DISCUSSION AND CONCLUSIONS: _____

NAME _____

DATE _____

SECTION _____

EXPERIMENT 4-11

Reduction of Nitrates

PURPOSE: _____

_____

_____

DATA:

Tube 1. Nitrate broth inoculated with Escherichia coli.

Day	Nitrite Production	Ammonia Production

Tube 2. Nitrate broth inoculated with Pseudomonas aeruginosa.

Day	Nitrite Production	Ammonia Production

Tube 3. Nitrate broth inoculated with Bacillus subtilis.

Day	Nitrite Production	Ammonia Production

DISCUSSION AND CONCLUSIONS: _____

_____

_____

_____

_____

_____

_____

_____

NAME _____

DATE _____

SECTION _____

EXPERIMENT 4-12

The Production of Hydrogen Sulfide by Microorganisms

PURPOSE: _____
_____

DATA:

Tube 1. Peptone iron agar inoculated with <u>Escherichia coli.</u>

Day	Observations

Tube 2. Peptone iron agar inoculated with <u>Proteus vulgaris.</u>

Day	Observations

DISCUSSION AND CONCLUSIONS:

NAME _____

DATE _____

SECTION _____

EXPERIMENT 4-13

Demonstration of the Production of Indole

PURPOSE: _____

_____

_____

DATA:

Tube 1. Tryptone broth inoculated with <u>Escherichia coli.</u>

_____

Tube 2. Tryptone broth inoculated with <u>Bacillus subtilis.</u>

_____

DISCUSSION AND CONCLUSIONS: _____

_____

_____

_____

_____

_____

_____

NAME _____

DATE _____

SECTION _____

EXPERIMENT 4-14

The MR-VP Procedure

PURPOSE: _____
_____
_____

DATA:

Microorganism	Methyl Red Test	Voges-Proskauer Test
Aerobacter aerogenes		
Escherichia coli		

DISCUSSION AND CONCLUSIONS: _____
_____
_____
_____
_____
_____
_____
_____

NAME _____

DATE _____

SECTION _____

EXPERIMENT 4-15

Detection of the Enzyme Catalase in Bacteria

PURPOSE: _____
_____

DATA:

Plate 1. <u>Escherichia coli</u> on tryptose phosphate agar with hydrogen peroxide added after incubation.

_____

Plate 2. <u>Lactobacillus delbrueckii</u> on tryptose phosphate agar with hydrogen peroxide added after incubation.

_____

DISCUSSION AND CONCLUSIONS: _____
_____
_____
_____
_____
_____
_____
_____

NAME _____

DATE _____

SECTION _____

EXPERIMENT 5-1

The Effect of Hydrogen Ion Concentration (pH) on Microorganisms

PURPOSE: _____

_____

DATA:

Escherichia coli in nutrient broth.

Day	pH 3.0	pH 5.0	pH 7.0	pH 9.0	pH 11.0
1					
2					
5					

Saccharomyces cerevisiae in malt extract broth.

Day	pH 3.0	pH 5.0	pH 7.0	pH 9.0	pH 11.0
1					
2					
5					

DISCUSSION AND CONCLUSIONS:

NAME _____

DATE _____

SECTION _____

EXPERIMENT 5-2

The Effect of Osmotic Pressure on Microorganisms

PURPOSE: _____
_____

DATA:

Escherichia coli in nutrient broth with glucose.

Day	0% (Normal)	5%	10%	25%

Aspergillus niger in malt extract broth with glucose.

Day	0% (Normal)	15%	30%	40%

DISCUSSION AND CONCLUSIONS: _____
_____
_____
_____

NAME _____

DATE _____

SECTION _____

EXPERIMENT 5-3

The Effect of Ultraviolet Radiation on Microorganisms

PURPOSE: _____
_____

DATA:

PLATE	RESULTS
Plate 1. Without the lid, exposed to ultraviolet light for 5 min.	
Plate 2. Without the lid, exposed to ultraviolet light for 10 min.	
Plate 3. With the lid, exposed to the ultraviolet light for 10 min.	

DISCUSSION AND CONCLUSIONS: _____
_____
_____
_____
_____
_____
_____

NAME _____

DATE _____

SECTION _____

EXPERIMENT 5-4

The Effect of Heat on Microbial Survival

PURPOSE: _____

_____

_____

_____

DATA:

Nutrient agar plate inoculated with Escherichia coli.

Temperature	Control	Uninoculated	5 min.	10 Min.	15 Min.	20 Min.
50°C.						
70°C.						
85°C.						

Nutrient agar plate inoculated with Bacillus subtilis.

Temperature	Control	Uninoculated	5 Min.	10 Min.	15 Min.	20 Min.
50°C.						
70°C.						
85°C.						

Nutrient agar plate inoculated with Staphylococcus aureus.

Temperature	Control	Uninoculated	5 Min.	10 Min.	15 Min.	20 Min.
50°C.						
70°C.						
85°C.						

DISCUSSION AND CONCLUSIONS:

NAME _____

DATE _____

SECTION _____

EXPERIMENT 5-5

Demonstration of Molecular Diffusion in Agar

PURPOSE: _____

_____

DATA:

A. 0.5% Methylene blue.
B. 0.5% Crystal violet.
C. 0.5% Safranin.

DISCUSSION AND CONCLUSIONS: _____

_____

_____

_____

_____

_____

_____

NAME _____

DATE _____

SECTION _____

EXPERIMENT 5-6

The Selective Action of Crystal Violet

PURPOSE: _____

_____

DATA:

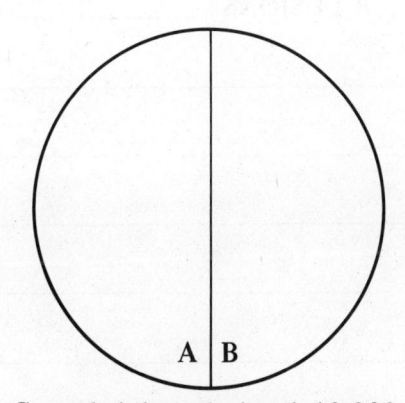

Crystal violet solution 1:10,000.
    A. Staphylococcus aureus.
    B. Escherichia coli.

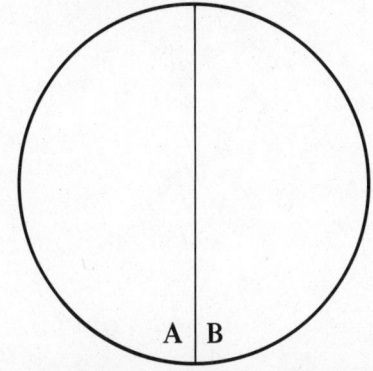

Crystal violet solution 1:50,000.
    A. Staphylococcus aureus.
    B. Escherichia coli.

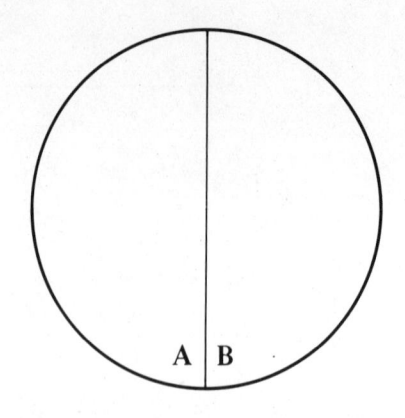

Crystal violet solution 1:100,000.
   A. <u>Staphylococcus aureus</u>.
   B. <u>Escherichia coli.</u>

DISCUSSION AND CONCLUSIONS: _____

_____

_____

_____

_____

_____

_____

NAME _____

DATE _____

SECTION _____

EXPERIMENT 5-7

The Oligodynamic Action of Heavy Metals

PURPOSE: _____
_____

DATA:

| Escherichia coli. | Staphylococcus aureus. |

DISCUSSION AND CONCLUSIONS: _____
_____
_____
_____
_____
_____
_____

NAME _____

DATE _____

SECTION _____

EXPERIMENT 5-8

The Filter Paper Disc Method of Evaluating Proprietary (Commercial) Antiseptics

PURPOSE: _____
_____

DATA:

Nutrient Agar

A	B
C	D

A	B
C	D

<u>Escherichia coli.</u>
A.
B.
C.
D.

<u>Staphylococcus aureus.</u>
A.
B.
C.
D.

245

Nutrient Agar + Peptone (or Blood Plasma)

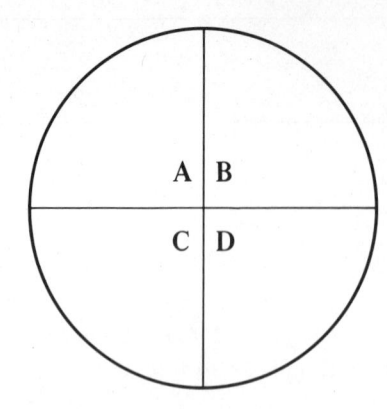

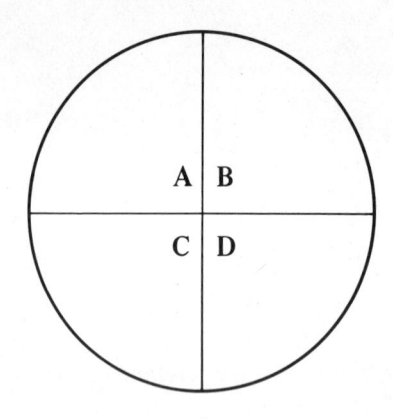

```
        A | B                    A | B
        C | D                    C | D
```

Escherichia coli.   Staphylococcus aureus.
A.   A.
B.   B.
C.   C.
D.   D.

DISCUSSION AND CONCLUSIONS: _____

_____
_____
_____
_____
_____
_____
_____
_____
_____
_____
_____

NAME _____

DATE _____

SECTION _____

EXPERIMENT 5-9

The Evaluation of Antibiotics by the Filter Paper Disc Method

PURPOSE: _____
_____

DATA:

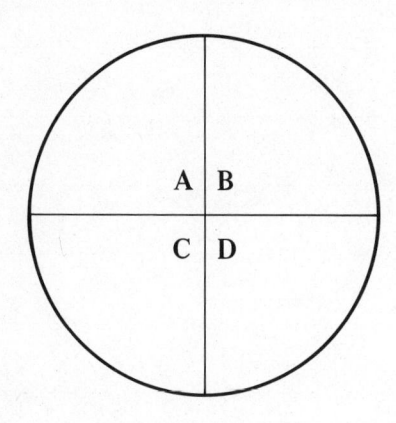

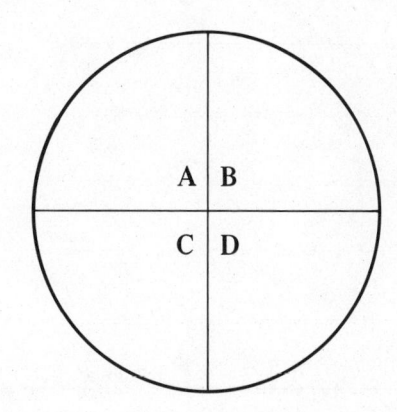

Staphylococcus aureus.
A. Penicillin
B. Streptomycin
C. Tetracycline
D. Chloromycetin

Escherichia coli.
A. Penicillin
B. Streptomycin
C. Tetracycline
D. Chloromycetin

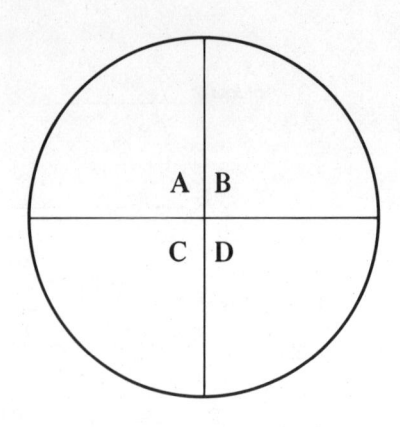

Micrococcus luteus.
A. Penicillin
B. Streptomycin
C. Tetracycline
D. Chloromycetin

DISCUSSION AND CONCLUSIONS:

NAME _____

DATE _____

SECTION _____

EXPERIMENT 5-10

Determination of the Rate of Action of Antiseptics with and without Organic Matter Present

PURPOSE: _____

_____

_____

DATA:

Nutrient broth inoculated with Staphylococcus aureus grown in nutrient broth.

Antiseptic	1 Min.	2 Min.	4 Min.	8 Min.

Peptone-rich nutrient broth inoculated with Staphylococcus aureus grown in peptone-rich nutrient broth.

Antiseptic	1 Min.	2 Min.	4 Min.	8 Min.

DISCUSSION AND CONCLUSIONS:

NAME _____

DATE _____

SECTION _____

EXPERIMENT 5-11

Some Observations on Skin Cleanliness and Skin Antisepsis

PURPOSE: _____
_____
_____

DATA

Sector	Results
1. Hands not washed.	
2. Hands washed once	
3. Hands washed two times.	
4. Hands washed three times.	
5. Swab from area treated with antiseptic for 2 min.	
6. Scrapings from area treated with antiseptic for 4 min.	

DISCUSSION AND CONCLUSIONS:

NAME _____

DATE _____

SECTION _____

EXPERIMENT 6-1

Identification of an Unknown Bacterium

Organism _____   Culture No. _____

Family _____   Natural Habitat _____

Gram _____   Morphology and Arrangement _____

Motile _____   Capsule _____   Spores _____

Pigment (color) _____   Acid Fast _____   Optimum Temp. ___

Hemolysis (type) _____   Pathogenic _____   Other _____

Gelatin Hydrolysis _____   Coagulase _____

Litmus Milk _____   Catalase _____

Nitrate Reduction _____   Starch Hydrolysis _____

Indole Produced _____

MR _____

VP _____

TSI (H_2S, gas, butt, slant) _____

Glucose _____ Special Tests:

Sucrose _____

Lactose _____

Maltose _____

Mannitol _____

Mannose _____

Fructose _____

Xylose _____

Rhamnose _____

Sorbitol _____

Inulin _____

Trehalose _____

Galactose _____

Arabinose _____

NAME _____

DATE _____

SECTION _____

EXPERIMENT 6-1

Identification of an Unknown Bacterium

Organism _____   Culture No. _____

Family _____   Natural Habitat _____

Gram _____   Morphology and Arrangement _____

Motile _____   Capsule _____   Spores _____

Pigment (color) _____   Acid Fast _____   Optimum Temp. ____

Hemolysis (type) _____   Pathogenic _____   Other _____

Gelatin Hydrolysis _____   Coagulase _____

Litmus Milk _____   Catalase _____

Nitrate Reduction _____   Starch Hydrolysis _____

Indole Produced _____

MR _____

VP _____

TSI (H_2 S, gas, butt, slant) _____

Glucose _____ Special Tests:

Sucrose _____

Lactose _____

Maltose _____

Mannitol _____

Mannose _____

Fructose _____

Xylose _____

Rhamnose _____

Sorbitol _____

Inulin _____

Trehalose _____

Galactose _____

Arabinose _____

NAME _____

DATE _____

SECTION _____

EXPERIMENT 6-1

Identification of an Unknown Bacterium

Organism _____ Culture No. _____

Family _____ Natural Habitat _____

Gram _____ Morphology and Arrangement _____

Motile _____ Capsule _____ Spores _____

Pigment (Color) _____ Acid Fast _____ Optimum Temp. ___

Hemolysis (type) _____ Pathogenic _____ Other _____

Gelatin Hydrolysis _____ Coagulase _____

Litmus Milk _____ Catalase _____

Nitrate Reduction _____ Starch Hydrolysis _____

Indole Produced _____

MR _____

VP _____

TSI (H_2S, gas, butt, slant) _____

Glucose _____   Special Tests:

Sucrose _____

Lactose _____

Maltose _____

Mannitol _____

Mannose _____

Fructose _____

Xylose _____

Rhamnose _____

Sorbitol _____

Inulin _____

Trehalose _____

Galactose _____

Arabinose _____

NAME _____

DATE _____

SECTION _____

EXPERIMENT 6-1

Identification of an Unknown Bacterium

Organism _____ Culture No. _____

Family _____ Natural Habitat _____

Gram _____ Morphology and Arrangement _____

Motile _____ Capsule _____ Spores _____

Pigment (color) _____ Acid Fast _____ Optimum Temp. ___

Hemolysis (type) _____ Pathogenic _____ Other _____

Gelatin Hydrolysis _____ Coagulase _____

Litmus Milk _____ Catalase _____

Nitrate Reduction _____ Starch Hydrolysis _____

Indole Produced _____

MR _____

VP _____

TSI (H_2S, gas, butt, slant) _____

Glucose _____ Special Tests:

Sucrose _____

Lactose _____

Maltose _____

Mannitol _____

Mannose _____

Fructose _____

Xylose _____

Rhamnose _____

Sorbitol _____

Inulin _____

Trehalose _____

Galactose _____

Arabinose _____

NAME _____

DATE _____

SECTION _____

EXPERIMENT 7-1

Bacterial Commensalism

PURPOSE: _____
_____

DATA:

Test Tube	Turbidity	Gram Stain
1. Staphylococcus aureus.		
2. Clostridium sporogenes.		
3. Staphylococcus aureus and Clostridium sporogenes.		

DISCUSSIONS AND CONCLUSIONS: _____
_____
_____
_____
_____
_____
_____

NAME _____

DATE _____

SECTION _____

EXPERIMENT 7-2

Experimental Antibiosis

PURPOSE: _____

_____

_____

DATA:

Escherichia coli in nutrient agar streaked with Bacillus subtilis.

Micrococcus luteus in nutrient agar streaked with Bacillus subtilis.

Organism	Antibiosis observed (+ or −)
Escherichia coli	
Micrococcus luteus	

DISCUSSION AND CONCLUSIONS: _____

_____

_____

_____

NAME _____

DATE _____

SECTION _____

EXPERIMENT 7-3

Antibiosis in Soil

PURPOSE: _____
_____

DATA:

Growth on Sabouraud agar plates (evidence of antibiosis).

Day	1:500	1:1000	1:2000
1			
2			

Type of organisms predominating. _____

Growth on nutrient agar plates (evidence of antibiosis).

Day	1:500	1:1000	1:2000
1			
2			

Type of organisms predominating. _____

ISOLATED COLONY (antibiotic-producing organism):

Type of organism (Gram stain, bacteria, yeast, mold, etc.). _____

_____

_____

Effective against what type of organisms? _____

_____

_____

DISCUSSION AND CONCLUSIONS: _____

NAME _____

DATE _____

SECTION _____

EXPERIMENT 7-4

Bacterial Synergism

PURPOSE: _____
_____
_____

DATA:

Lactose Broth.

Tube	Results
1. Staphylococcus aureus.	
2. Proteus vulgaris.	
3. S. aureus and P. vulgaris.	

Sucrose Broth.

Tube	Results
1. Staphylococcus aureus.	
2. Proteus vulgaris.	
3. S. aureus and P. vulgaris.	

DISCUSSION AND CONCLUSIONS: _____
_____
_____
_____

NAME _____

DATE _____

SECTION _____

EXPERIMENT 7-5

Symbiotic Nitrogen-Fixing Bacteria

PURPOSE: _____
_____
_____

DATA:

1. Nodule smear.
2. Methylene blue.
3. Oil-immersion.

1. Smear from Rhizobium colony.
2. Gram stain.
3. Oil-immersion.

DISCUSSION AND CONCLUSIONS: _____
_____
_____
_____
_____
_____

NAME _____

DATE _____

SECTION _____

EXPERIMENT 7-6

The Relationship of Microorganisms to Dental Decay

PURPOSE: _____

_____

_____

DATA:

Incubation	Results from Test Tube with Saliva Sample
24 Hours	
48 Hours	
72 Hours	

DISCUSSION AND CONCLUSIONS: _____

_____

_____

_____

_____

_____

_____

_____

NAME _____

DATE _____

SECTION _____

EXPERIMENT 7-7

Normal Flora of the Human Skin

PURPOSE: _____

_____

DATA:

	Colony 1	Colony 2
Description		
Gram Stain		
Motility		
Spore Stain		
Glucose		
Sucrose		
Lactose		
Hydrogen Sulfide		
Probable Species		

DISCUSSION AND CONCLUSIONS: _____

_____

_____

_____

NAME _____

DATE _____

SECTION _____

EXPERIMENT 7-8

Normal Flora of the Human Throat

PURPOSE: _____
_____

DATA:

	Colony 1	Colony 2	Colony 3
Description			
Hemolysis			
Gram Stain			
Motility			
Trehalose			
Glucose			
Sucrose			
Lactose			
Inulin			
Mannitol			
Possible Species			

DISCUSSION AND CONCLUSIONS:

NAME _____

DATE _____

SECTION _____

EXPERIMENT 8-1

The Bacteriological Analysis of Water

A. Presumptive Test

PURPOSE: _____
_____
_____

DATA:

Water Sample Code Number	24 Hours Acid or Gas in Lactose	48 Hours Acid or Gas in Lactose

B. Confirmed Test

PURPOSE: _____

_____

_____

DATA:

Water Sample Code Number	EMB Agar (Pigment)	Gram Stain

C. Completed Test

PURPOSE: _____

_____

_____

DATA:

Water Sample Code No.				
Lactose				
Tryptone (Indol)				
MR-VP				
Citrate				
Gram Stain				
Identify Organism				

DISCUSSION AND CONCLUSIONS: _____

_____

_____

NAME _____

DATE _____

SECTION _____

EXPERIMENT 8-2

The Bacterial Growth Curve

PURPOSE: _____

_____

_____

DATA:

	Number of Bacterial Colonies in Petri Dish					
	0 hours		1 hour		2 hours	
	Group	Class Average	Group	Class Average	Group	Class Average
Diluted $(10)^{-5}$ X						
Diluted $(10)^{-7}$ X						
Average number of bacterial/ml. of original culture						

Growth Curve

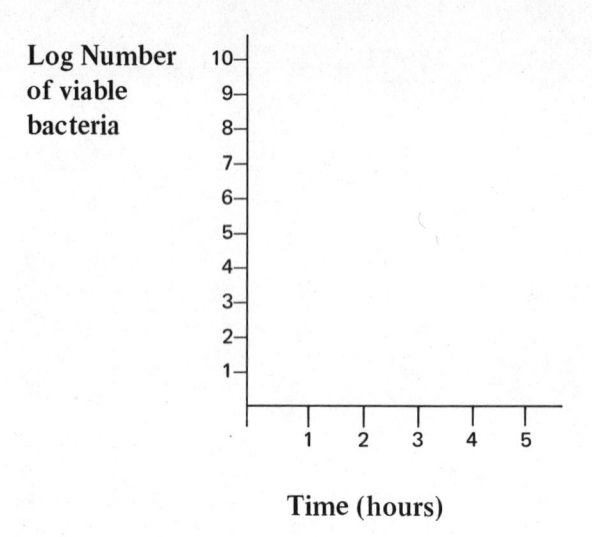

Log Number of viable bacteria

(y-axis: 1–10; x-axis: Time (hours) 1–5)

DISCUSSION AND CONCLUSIONS: _____

NAME _____

DATE _____

SECTION _____

EXPERIMENT 8-3

The Plate Count of Bacteria in Milk

PURPOSE: _____
_____
_____

DATA:

Bacterial Counts in Raw Milk

	Number of bacterial colonies per dish		No. of cells/ml. of original milk sample	
Dilution	Group	Class Average	Group	Class
1:10				
1:100				
1:1000				
1:10,000				
	Final Average All Dilutions (bacteria/ml.)			

Bacterial Counts in Pasteurized Milk

	Number of Bacterial Colonies per Dish		No. of cells/ml. of original milk sample	
Dilution	Group	Class Average	Group	Class
1:10				
1:100				
1:1000				
		Final Average All Dilutions (Bacteria/ml.)		

DISCUSSION AND CONCLUSIONS: _____

_____
_____
_____
_____
_____
_____
_____

NAME _____

DATE _____

SECTION _____

EXPERIMENT 8-4

The Effect of Temperature on Microbial Growth in Milk

PURPOSE: _____

_____

_____

DATA:

Incubation of raw milk at 37°C. for 48 hours.

Odor _____

Litmus Milk at 37°C.

Day 1 _____

Day 2 _____

Day 3 _____

Day 4 _____

Day 5 _____

Day 6 _____

Gram stain. Oil immersion.
37°C./48 hrs.

Day 7 _____

Incubation of raw milk at 20°C. for 4-5 days.

Odor _____

Litmus Milk at 20°C.

Day 1 _____

Day 2 _____

Day 3 _____

Day 4 _____

Day 5 _____

Day 6 _____

Gram stain. Oil immersion.
20°C./4-5 days.

Day 7 _____

DISCUSSION AND CONCLUSIONS: _____

_____

_____

_____

_____

_____

_____

_____

NAME _____

DATE _____

SECTION _____

EXPERIMENT 9-2

Selective Medium for the Isolation of Pathogenic Staphylococci

PURPOSE: _____

_____

DATA:

	Staph. aureus	M. luteus	E. coli	Throat Swab
Colony Color and Size				
Zone Color				
Growth				
Description and Gram Stain of Each Microorganism				

DISCUSSION AND CONCLUSIONS: _____

_____

_____

_____

_____

_____

NAME _____

DATE _____

SECTION _____

EXPERIMENT 9-3

The Agglutination Reaction

PURPOSE: _____

_____

_____

DATA:

Organism (Antigen)	Tube Number									
	1	2	3	4	5	6	7	8	9	10
Staphylococcus aureus										
Escherichia coli										

Titer of the Staph. aureus antiserum _____.

Titer of the E. coli antiserum _____.

DISCUSSION AND CONCLUSIONS: _____

_____

_____

_____

_____

_____

_____

NAME _____

DATE _____

SECTION _____

EXPERIMENT 9-4

The Morphology of Blood Cells, Erythrocyte Blood Typing and Differential Count

DATA:

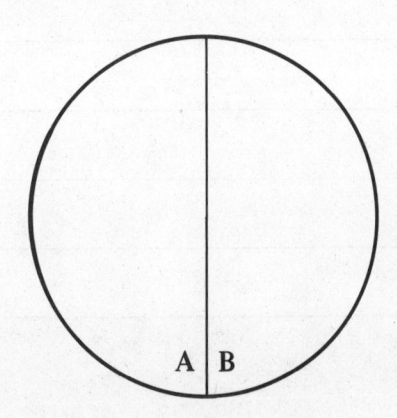

Blood Type _____

1. Erythrocyte agglutination.
2. Wet mount.
3. Low power.

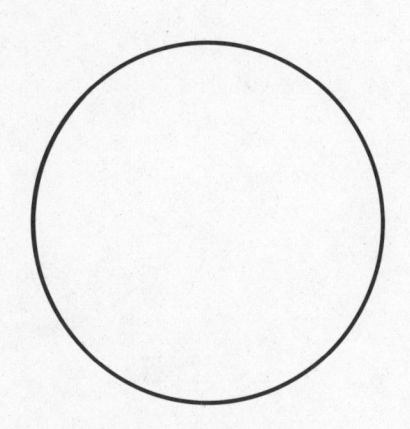

1. Blood smear.
2. Wright's stain.
3. Oil-immersion.

289

Differential White Blood Cell Count				
Neutrophils	Basophils	Eosinophils	Lymphocytes	Monocytes
%	%	%	%	%

DISCUSSION AND CONCLUSIONS: _____

_____

_____

_____

_____

_____

_____

APPENDIX A

ORGANISMS USED FOR EXPERIMENTS IN THIS BOOK**

1. Aerobacter aerogenes
2. Aerobacter cloacae*
3. Alcaligenes bookeri*
4. Alcaligenes faecalis
5. Alcaligenes marshallii*
6. Alcaligenes metacaligenes*
7. Alcaligenes recti*
8. Alcaligenes viscolactis*
9. Aspergillus niger
10. Bacillus cereus*
11. Bacillus megaterium*
12. Bacillus polymyxa*
13. Bacillus subtilis
14. Clostridium sporogenes
15. Corynebacterium pseudodiphtheriticum
16. Corynebacterium xerosis*
17. Diplococcus pneumoniae
18. Erwinia amylovora*
19. Escherichia aurescens*
20. Escherichia coli
21. Escherichia freundii*
22. Gaffkya tetragena*
23. Klebsiella pneumoniae
24. Lactobacillus casei*
25. Lactobacillus delbruekii
26. Lactobacillus lactis*
27. Micrococcus luteus
28. Micrococcus roseus*
29. Micrococcus rubens*
30. Micrococcus ureae*
31. Mycobacterium tuberculosis (avirulent strain)
32. Neisseria catarrhalis*
33. Neisseria flava*
34. Neisseria perflava*
35. Penicillium notatum
36. Proteus mirabilis*
37. Proteus morganii*
38. Proteus vulgaris
39. Pseudomonas aeruginosa
40. Pseudomonas fluorescens*
41. Saccharomyces cerevisiae
42. Salmonella cholerasius*
43. Salmonella schottmuelleri*
44. Salmonella typhimurium*

*These organisms are used only in the identification of unknown bacteria experiments. Not all of them would have to be kept in stock.

**Several of the organisms listed here from the Enterobacteriaceae have acquired changed names and subclassifications according to Edwards and Ewing, "Identification of Enterobacteriaceae" (1972). However, in the absence of a new edition of Bergey's Manual beyond 1958, this book will continue to employ the classification used in Bergey's in order to avoid any confusion. Hopefully, the new edition of Bergey's Manual will be available by the next revision of this text.

5. Ammonium Phosphate Broth

Ammonium Phosphate	
($NH_4H_2PO_4$)	1.0 g.
Dextrose (glucose)	10.0 g.
Potassium Chloride	
(KCl)	0.2 g.
Magnesium Sulfate	
($MgSO_4$)	0.2 g.
Bromcresol Purple	
(1.6%)	1.0 ml.

Make up the bromcresol purple fresh. Then add the above ingredients to one liter of distilled water. Stir to completely dissolve and dispense. Final pH should be 7.0. Autoclave at 15 p.s.i. pressure for 15 min.

6. Blood Agar

(a) Make up the desired amount of Nutrient Agar (with 0.8% NaCl added) with 5 per cent decrease in the distilled water used, or use Blood Agar Base (commercially available).

(b) Autoclave, then cool to 60°C. and add 5 ml. of blood (rabbit or horse) per 95 ml. of agar used. Pour into sterile Petri dishes. Flame the top of the agar to remove air bubbles.

7. Calcium Chloride Agar

Nutrient Agar	1000.0 ml.
Calcium Chloride	
($CaCl_2$)	0.5 g.

When making up the nutrient agar, add 0.5 g. of calcium chloride. Heat in covered double boiler until completely dissolved. Dispense and autoclave at 15 p.s.i. pressure for 15 min.

8. E.M.B. Agar

E.M.B. Agar (purchased	
commercially)	36.0 g.
or	
Peptone	10.0 g.
Lactose	5.0 g.
Saccharose	
(sucrose)	5.0 g.
Dipotassium Phosphate	
(K_2HPO_4)	2.0 g.
Agar	13.5 g.
Eosin	0.4 g.
Methylene Blue	0.065 g.

Add the above ingredients to one liter of distilled water. Heat in covered double boiler until completely dissolved. Dispense and autoclave at 15 p.s.i. pressure for 15 min.

9. Fermentation Broths

Fermentable Substance	
Desired	7.0 g.
Phenol Red Broth Base	
(commercially	
available)	16.0 g.
or	
Beef Extract	1.0 g.
Peptone	10.0 g.
Sodium Chloride	
(NaCl)	5.0 g.
Phenol Red	0.018 g.

Add the above ingredients to one liter of distilled water. Stir to completely dissolve. Add a Durham tube to each of the test tubes. Dispense the broth and then autoclave at 10 p.s.i. pressure for 20 min. (Many of the fermentation broths cannot be autoclaved at 10 p.s.i. pressure; there are others that cannot be autoclaved at 7 p.s.i. pressure without damage. A few fermentation broths must be sterilized by putting them through a bacteriological filter.

Glucose, lactose and sucrose can be autoclaved at 10 p.s.i. pressure. Fructose and xylose must be sterilized by utilizing a bacteriological filter. All other fermentation broths listed in the chart Specific Characteristics of Selected Bacteria should be autoclaved at 7 p.s.i. pressure.)

10. **Glycerol Agar**

Nutrient Agar 1000.0 ml.
Glycerine
 (glycerol) 50.0 ml.
Make up one liter of Nutrient Agar, but with 5% less water than normal. Add 5 ml. of glycerine per 95 ml. of agar used. Stir to mix thoroughly. Dispense and autoclave at 15 p.s.i. pressure for 15 min.

11. **Hay Infusion**

Put a handful of organic debris, such as grass clippings, straw and old leaves, in a jar of tap water. Insert an air hose to the bottom of the jar and gently bubble air through the water. It often helps to add a teaspoon of dried nutrient broth powder to the infusion as well. Incubate at room temperature with constant aeration for about five days, replenishing water as it evaporates.

12. **Litmus Milk**

Litmus Milk Medium
 (commercially
 available) 105.0 g.
 or
Skim Milk 100.0 g.
Litmus 0.75 g.
Add the above ingredients to one liter of distilled water and stir to completely dissolve. Dispense and autoclave at 7 p.s.i. pressure for 20 min. The best way to sterilize litmus milk is by boiling in a water bath for 20 min. on 3 successive days, since milk tends to partially hydrolyse at higher temperatures. *All* milk-containing media should be treated in this fashion.

13. **Malt Extract Agar**

Malt Extract 15.0 g.
Agar 15.0 g.
Add the above ingredients to one liter of distilled water. Heat in covered double boiler until completely dissolved. After dispensing, autoclave at 15 p.s.i. pressure for 15 min.

14. **Malt Extract Broth**

Malt Extract 15.0 g.
Add the above to one liter of distilled water. Stir to completely dissolve. Dispense and autoclave at 15 p.s.i. pressure for 15 min.

15. **Mannitol Salt Agar**

Mannitol Salt Agar
 (commercially
 available) 111.0 g.
 or
Beef extract 1.0 g.
Peptone
 (proteose) 10.0 g.
Sodium Chloride
 (NaCl) 75.0 g.
Mannitol 10.0 g.
Agar 15.0 g.
Phenol Red 0.025 g.
Add the above ingredients to one liter of distilled water. Heat in covered double boiler until

completely dissolved. Dispense and autoclave at 15 p.s.i. for 15 min.

16. **M.R.-V.P. Broth (Methyl Red-Voges-Proskauer Test)**

M.R.-V.P. Medium (commercially available)	17.0 g.

or

Buffered Peptone	7.0 g.
Dextrose (glucose)	5.0 g.
Dipotassium Phosphate (K_2HPO_4)	5.0 g.

Dissolve the above in one liter of distilled water. Stir to completely dissolve. Dispense and autoclave at 15 p.s.i. pressure for 15 min.

17. **Nitrate Broth**

Potassium Nitrate (KNO_3)	1.0 g.
Sodium Chloride (NaCl)	0.5 g.
Peptone	2.0 g.

Add the above ingredients to one liter of distilled water. Stir to completely dissolve. Dispense after adding one Durham tube to each test tube. Autoclave at 15 p.s.i. pressure for 15 min.

18. **Nutrient Agar**

Nutrient Agar (commercially available)	23.0 g.

or

Beef Extract	3.0 g.
Peptone	5.0 g.
Agar	15.0 g.

Add the above ingredients to one liter of distilled water. Heat in covered double boiler until completely dissolved. Dispense and autoclave at 15 p.s.i. pressure for 15 min.

19. **Nutrient Broth**

Nutrient Broth (commercially available)	8.0 g.

or

Beef Extract	5.0 g.
Peptone	3.0 g.

Add the above ingredients to one liter of distilled water. Stir until completely dissolved. Dispense and autoclave at 15 p.s.i. pressure for 15 min.

20. **Nutrient Gelatin**

Nutrient Gelatin Medium (commercially available)	128.0 g.

or

Beef Extract	3.0 g.
Peptone	5.0 g.
Gelatin	120.0 g.

Add the above ingredients to one liter of distilled water. Heat in covered double boiler until completely dissolved. Dispense and autoclave at 7 p.s.i. pressure for 20 min.

21. **Nutrient Gelatin (0.4%) Agar**

Tryptose	20.0 g.
Yeast Extract	0.3 g.
Manganese Sulfate ($MnSO_4$)	0.1 g.
Gelatin	4.0 g.
Agar	15.0 g.

Add the above ingredients to one liter of distilled water. Heat in covered double boiler until completely dissolved. Dispense and autoclave at 10 p.s.i. pressure for 20 min.

22. Peptone-Rich Nutrient Broth

Peptone 92.0 g.
Nutrient Broth 8.0 g.
Add the above ingredients to one liter of distilled water. Stir until completely dissolved. Dispense and autoclave at 15 p.s.i. pressure for 15 min.

23. Peptone-Iron Agar

Peptone-Iron Agar
  (commercially
  available) 36.0 g.
      or
Peptone 15.0 g.
Proteose-Peptone 5.0 g.
Ferric Ammonium
  Citrate 0.5 g.
Sodium Thiosulfate
  ($Na_2S_2O_3 \cdot 5H_2O$) 0.08 g.
Dipotassium Phosphate
  (K_2HPO_4) 1.0 g.
Agar 15.0 g.
Add the above ingredients to one liter of distilled water. Heat in a covered double boiler until completely dissolved. Adjust pH to 6.7 and dispense. Autoclave at 15 p.s.i. pressure for 15 min.

24. Peptone Solution (8%) (Protein-like)

Peptone 8.0 g.
Add the peptone to 100 ml. of distilled water. Stir until completely dissolved. Dispense and autoclave at 15 p.s.i. pressure for 15 min.

25. Peptone Solution (10%)

Peptone 10.0 g.
Add the peptone to 100 ml. of distilled water. Stir until completely dissolved. Dispense and autoclave at 15 p.s.i. pressure for 15 min.

26. Phenol Agar (0.3% Phenol)

Nutrient Agar 1000.0 ml.
Phenol Crystals, C.P. 3.0 g.
When making up the nutrient agar, add 3.0 g. of phenol crystals. Heat in covered double boiler to completely dissolve ingredients. Dispense and autoclave at 15 p.s.i. pressure for 15 min.

27. Phenol Red Agar

Dextrose 5.0 g.
Phenol Red Agar Base
  (commercially
  available) 30.0 g.
      or
Beef Extract 1.0 g.
Proteose Peptone 10.0 g.
Sodium Chloride
  (NaCl) 5.0 g.
Agar 15.0 g.
Phenol Red 0.025 g.
Add the above ingredients to one liter of distilled water. Heat in covered double boiler until completely dissolved. Dispense and autoclave at 15 p.s.i. pressure for 15 min.

28. Sabouraud (Maltose) Agar

Neopeptone 10.0 g.
Maltose 40.0 g.
Agar 15.0 g.
Add the above ingredients to one liter of distilled water. Heat in covered double boiler until completely dissolved. Adjust pH to 5.6 and dispense. Autoclave at 15 p.s.i. pressure for 15 min.

29. **Skim Milk**

   (a) Procure the desired amount of fresh skim milk.
   (b) Dispense 7.5 ml. into each test tube.
   (c) Boil for 3 successive days. Autoclaving will *not* do.

30. **Snyder Test Agar (Trypsin-Digest Agar)**

Snyder Test Agar (commercially available)	65.0 g.
or	
Tryptone	20.0 g.
Dextrose (glucose)	20.0 g.
Sodium Chloride (NaCl)	5.0 g.
Agar	20.0 g.
Bromcresol Green	0.02 g.

   Add the above ingredients to one liter of distilled water. Heat in covered double boiler until completely dissolved. Adjust pH to 4.8 and dispense. Autoclave at 15 p.s.i. pressure for 15 min.

31. **Sodium Chloride Agar (7% Sodium Chloride)**

Sodium Chloride (NaCl)	70.0 g.
Nutrient Agar	1000.0 ml.

   When making up the nutrient agar, add the 70.0 g. of sodium chloride. Heat in covered double boiler to completely dissolve ingredients. Dispense and autoclave at 15 p.s.i. pressure for 15 min.

32. **Sodium Citrate Broth**

Koser Citrate Medium (commercially available)	5.7 g.
or	
Sodium Ammonium Phosphate ($NaNH_4HPO_4$)	1.5 g.
Monopotassium Phosphate (KH_2PO_4)	1.0 g.
Magnesium Sulfate ($MgSO_4$)	0.2 g.
Sodium Citrate ($Na_3C_6H_5O_7$)	3.0 g.

   Dissolve the above ingredients in one liter of distilled water. Stir to completely dissolve. Adjust pH to 6.7 and dispense. Autoclave at 15 p.s.i. pressure for 15 min.

33. **Sodium Ricinoleate Agar**

Nutrient Agar	1000.0 ml.
Sodium Ricinoleate	1.0 g.

   When making up the nutrient agar, add the 1.0 g. of sodium ricinoleate. Heat in covered double boiler until ingredients are completely dissolved. Dispense and autoclave at 15 p.s.i. pressure for 15 min.

34. **S.S. Agar (Salmonella-Shigella Agar)**

S.S. Agar (commercially available)	60.0 g.
or	
Beef Extract	5.0 g.
Proteose Peptone	5.0 g.
Lactose	10.0 g.
Bile Salts	8.5 g.
Sodium Citrate ($Na_3C_6H_5O_7$)	8.5 g.
Sodium Thiosulfate ($Na_2S_2O_3$)	8.5 g.
Ferric Citrate ($FeC_6H_5O_7$)	1.0 g.
Agar	13.5 g.
Brilliant Green	0.00033 g.
Neutral Red	0.025 g.

   Add the above ingredients to one liter of distilled water. Heat in covered double boiler until completely dissolved. Adjust pH to 7.0. Do not autoclave.

35. **Standard Methods Agar (Plate Count Agar)**

   Plate Count Agar
     (commercially
     available)   23.5 g.
               or
   Yeast Extract   2.5 g.
   Tryptone   5.0 g.
   Dextrose (glucose)   1.0 g.
   Agar   15.0 g.
   Dissolve the above ingredients in one liter of distilled water. Heat to boiling to dissolve and autoclave for 15 min. at 15 p.s.i. to sterilize.

36. **Starch Agar**

   Nutrient Agar   1000.0 ml.
   Starch (soluble)   2.0 g.
   When making up the nutrient agar, add 2.0 g. of starch to the dry ingredients. Heat in covered double boiler until the ingredients are completely dissolved. Adjust pH to 6.8-7.0 and dispense. Autoclave at 15 p.s.i. pressure for 15 min.

37. **Thioglycollate Broth**

   Thioglycollate Medium
     (commercially
     available)   29.3   g.
               or
   Casitone   15.0   g.
   Yeast Extract   5.0   g.
   Dextrose (glucose)   5.0   g.
   Sodium Chloride
     (NaCl)   2.5   g.
   l-Cystine   0.5   g.
   Thioglycollic Acid   0.3   ml.
   Agar   0.75   g.
   Resazurin (certified)   0.001   g.
   Dissolve the above ingredients in one liter of distilled water. Adjust pH to 7.1 and dispense. Autoclave at 15 p.s.i. pressure for 15 min. Store in dark place.

38. **T.S.I. Agar (Triple Sugar Iron Agar)**

   T.S.I. Agar (commercially
     available)   65.0   g.
               or
   Beef Extract   3.0   g.
   Yeast Extract   3.0   g.
   Peptone   15.0   g.
   Proteose-Peptone   5.0   g.
   Lactose   10.0   g.
   Saccharose (sucrose)   10.0   g.
   Dextrose (glucose)   1.0   g.
   Ferrous Sulfate
     ($FeSO_4$)   0.2   g.
   Sodium Chloride
     (NaCl)   5.0   g.
   Sodium Thiosulfate
     ($Na_2S_2O_3$)   0.3   g.
   Agar   12.0   g.
   Phenol Red   0.024 g.
   Add the above ingredients to one liter of distilled water and heat in covered double boiler until completely dissolved. Adjust pH to 7.4 and then dispense. Autoclave at 15 p.s.i. pressure for 15 min.

39. **Tryptone Broth**

   Tryptone   10.0 g.
   Dissolve the above in one liter of distilled water. Autoclave at 15 p.s.i. pressure for 15 min.

40. **Tryptose Phosphate Agar**

   Tryptose Phosphate Agar
     (commercially
     available)   29.5 g.
               or
   Tryptose   20.0 g.
   Dextrose (glucose)   2.0 g.

Sodium Chloride
  (NaCl)   5.0 g.
Disodium Phosphate
  (Na_2HPO_4)   2.5 g.
Agar   15.0 g.

Add the above ingredients to one liter of distilled water and heat in covered double boiler to dissolve. Adjust pH to 7.3, dispense, and autoclave at 15 p.s.i. pressure for 15 min.

41. Urea Broth

Urea Broth (commercially
  available)   38.7 g.
            or
Yeast Extract   0.1 g.
Monopotassium Phosphate
  (KH_2PO_4)   9.1 g.
Disodium Phosphate
  (Na_2HPO_4)   9.5 g.
Urea   20.0 g.
Phenol Red   0.01 g.

Dissolve the above ingredients in one liter of distilled water. Adjust pH to 6.8, and after sterilizing with a bacteriological filter, dispense into sterile tubes.

42. Uric Acid Broth

Sodium Chloride
  (NaCl)   5.0 g.
Magnesium Sulfate
  ($MgSO_4 \cdot 7H_2O$) 0.2 g.
Calcium Chloride
  ($CaCl_2$)   0.1 g.
Dipotassium Phosphate
  (K_2HPO_4)   1.0 g.
Glucose (dextrose)   32.0 g.
Uric Acid   0.5 g.

Warm the above ingredients in one liter of distilled water in covered double boiler until dissolved. Dispense and autoclave at 15 p.s.i. pressure for 15 min.

43. Yeast Extract Mannitol Agar

Yeast Extract   5 g.
Mannitol   12 g.
K_2HPO_4 0.5 g.
$MgSO_4 \cdot 7H_2O$ 0.2 g.
$CaCl_2$ 0.2 g.
NaCl   0.1 g.
Agar   15 g.

Dissolve above ingredients in one liter of distilled water and heat to boiling to dissolve. Adjust pH to 7.0-7.2. Autoclave at 15 p.s.i. for 15 min. to sterilize.

APPENDIX C

REAGENTS EMPLOYED IN THIS BOOK

1. **Acetone-Alcohol**

   Ethanol (95%) 700.0 ml.
   Acetone 300.0 ml.
   Mix the two liquids.

2. **Acid-Alcohol**

   Hydrochloric Acid (37%),
     C.D. 30.0 ml.
   Ethanol (95%) 970.0 ml.
   Dissolve the hydrochloric acid in the ethanol.

3. **α-Naphthol Solution**

   α-Naphthol 5.0 g.
   Ethanol (95%) 100.0 ml.
   Dissolve the α-naphthol in the ethanol. Stir to mix thoroughly.

4. **Dimethyl-α-naphthylamine Solution**

   Dimethyl-α-
     naphthylamine 6.0 ml.
   Acetic Acid (5N) 1000.0 ml.
         or
   Glacial Acetic
     Acid (99%) 294.0 ml.
   Distilled Water 706.0 ml.
   Add the acetic acid to the distilled water, if making the 5N acetic acid first. Mix this solution and then add to the dimethyl-α-naphthylamine. Warm in a water bath to completely dissolve.

5. **Dipotassium Phosphate Buffer Solution (0.1 M)**

   Dipotassium Phosphate
     (K_2HPO_4) 22.8 g.
   Distilled Water 1000.0 ml.
   Dissolve the dipotassium phosphate in the distilled water.

6. **Egg Albumin (1.0%)**

   Dissolve 1 g. of lyophilized egg albumin in 100 ml. of physiological saline solution.

7. **Ethanol (10%)**

   Ethanol (95%) 52.0 ml.
   Add enough distilled water to make a total volume of 500 ml.

8. **Ethanol (40%)**

   Ethanol (95%) 210.0 ml.
   Add enough distilled water to make a total volume of 500 ml.

9. **Ethanol (70%)**

   Ethanol (95%) 368.4 ml.
   Add enough distilled water to make a total volume of 500 ml.

10. **1.0 N Hydrochloric Acid – 95% Alcohol Solution**

    Hydrochloric
      Acid (37%) 83.5 ml.
    Ethanol (95%) 916.5 ml.
    Dissolve the hydrochloric acid in the ethanol.

11. **Hydrochloric Acid (0.1 N)**

    Hydrochloric Acid
      (37%), C.P. 8.4 ml.
    Dissolve the hydrochloric acid in sufficient distilled water to make 1000 ml. total volume.

12. **Kovac's Solution**

    Para-dimethyl-amino-
      benzaldehyde 5.0 g.
    Amyl or Butyl
      Alcohol 75.0 ml.
    Hydrochloric Acid
      (37%), C.P. 25.0 ml.
    Dissolve the para-dimethyl-amino-benzaldehyde in the alcohol. Warm the solution gently using a water bath. After the ingredients are dissolved, carefully add the hydrochloric acid and stir.

13. **Methyl Cellulose Solution (10%)**

    Methyl cellulose
      powder 10 g.
    Tap water 100 ml.
    Heat the water to 85°C. (not boiling) and disperse the methyl cellulose powder in the water. Stir rapidly and constantly while cooling the mixture in an ice bath to about 5°C. The solution is now stable at room temperature and can be stored in a tightly closed container.

14. **Monopotassium Phosphate Buffer Solution (0.1 M)**

    Monopotassium Phosphate
      (KH_2PO_4) 13.6 g.
    Distilled Water 1000.0 ml.
    Dissolve the monopotassium phosphate in the distilled water.

15. **Nessler's Reagent**

    Dissolve 50 g. of potassium iodide in 35 ml. of cold distilled water. Add a saturated solution of mercuric chloride until a slight precipitate persists. Add 400 ml. of a 50% solution of potassium hydroxide. Dilute to one liter, allow to settle, and decant the supernatant for use.

16. **Phosphomolybdic Acid (1%)**

    Phosphomolybdic
      Acid 5.0 g.
    Distilled Water 500.0 ml.
    Dissolve the phosphomolybdic acid in the distilled water.

17. **Potassium Hydroxide Solution (M.R.-V.P. Test)**

    Potassium Hydroxide,
      C.P. (KOH) 40.0 g.
    Creatine 0.3 g.
    Distilled Water 100.0 ml.
    Dissolve the potassium hydroxide in 75 ml. of distilled water, and allow the solution to cool to room temperature. Then add the creatine and stir to dissolve. Add the remaining distilled water.

18. **Potassium Hydroxide 0.1 M.**

    Potassium Hydroxide
       (KOH)   5.6 g.
    Distilled Water   1000.0 ml.
    Add the potassium hydroxide to the distilled water. Mix until thoroughly dissolved.

19. **Saline Solution (0.9%)**

    Sodium Chloride
       (NaCl)   4.5 g.
    Distilled Water   500.0 ml.
    Dissolve the salt in the distilled water. Autoclave at 15 p.s.i. pressure for 15 min. to sterilize.

20. **Sodium Hydroxide (0.1 N)**

    Sodium Hydroxide,
       C.P. (NaOH)   0.4 g.
    Dissolve the sodium hydroxide in 100 ml. of distilled water.

21. **Sodium Hydroxide (4% or 1.0 N)**

    Sodium Hydroxide,
       C.P. (NaOH)   4.0 g.
    Dissolve the sodium hydroxide in 100 ml. of distilled water.

22. **Sodium Phosphate, Dibasic (0.1 M)**

    Sodium Phosphate, Dibasic
       (Na_2HPO_4)   14.2 g.
    Add the dibasic sodium phosphate to one liter of distilled water. Mix until dissolved.

23. **Sodium Phosphate, Monobasic (0.1 M)**

    Sodium Phosphate,
       Monobasic
       (NaH_2PO_4)   12.0 g.
    Add the monobasic sodium phosphate to one liter of distilled water. Mix until dissolved.

24. **Sulfanilic Acid**

    Sulfanilic Acid   8.0 g.
    Acetic Acid
       (5 N)   1000.0 ml.
                 or
    Glacial Acetic
       Acid   294.0 ml.
    Distilled Water   706.0 ml.
    Add the glacial acetic acid to the distilled water, if making the 5 N acetic acid first. Mix, and then add the sulfanilic acid and stir again.

APPENDIX D

STAINS EMPLOYED IN THIS BOOK

1. **Albert's Diphtheria Stain**

Toluidine Blue	0.15 g.
Methyl Green	0.20 g.
Acetic Acid (Glacial), C.P. (99%)	1.00 ml.
Ethanol (95%)	2.00 ml.
Distilled Water	100.00 ml.

   Dissolve the toluidine blue and the methyl green in the distilled water. Then add the acetic acid and the ethanol. Mix well.

2. **Bromcresol Purple (1.6%)**

Bromcresol Purple	16 g.
Distilled Water	500 ml.
Ethanol (95%)	500 ml.

   Dissolve the bromcresol purple in the ethanol, then add the distilled water.

3. **Crystal Violet 0.5%**

   Crystal Violet   1.25 g.
   Dissolve the Crystal Violet in 250 ml. of distilled water.

4. **Crystal Violet 1:2,000**

   Crystal Violet   0.5 g.
   Dissolve the Crystal Violet in one liter of distilled water.

5. **Crystal Violet 1:10,000**

   Make up a 1:2,000 solution of Crystal Violet. To 4 parts distilled water, add 1 part of the 1:2,000 solution.

6. **Crystal Violet 1:50,000**

   After making up the 1:10,000 solution of Crystal Violet, add to 4 parts distilled water one part of the 1:10,000 solution.

7. **Crystal Violet 1:100,000**

   After making up the 1:50,000 solution of Crystal Violet, add to 1 part distilled water one part of the 1:50,000 solution.

8. **Dorner's Nigrosin**

   Nigrosin   10.0 g.
   Add the Nigrosin to 100 ml. of distilled water. Boil 30 minutes. Add as a preservative 0.5 ml. of Formaldehyde (40%). Filter twice through filter paper and then store under aseptic conditions.

9. Gram's Crystal Violet

Solution A:
Crystal Violet   2.0 g.
Ethanol (95%)   20.0 ml.
Dissolve the Crystal Violet in the Ethanol.

Solution B:
Ammonium Oxalate,
  C.P.   0.8 g.
Distilled Water   80.0 ml.
Dissolve the Ammonium Oxalate in the distilled water.

After making up the two solutions, pour them together and stir until well mixed.

10. Gram's Iodine

Iodine, C.P.   1.0 g.
Potassium Iodide,
  C.P. (KI)   2.0 g.
Distilled Water   300.0 ml.
Mix the Iodine and the Potassium Iodide in a mortar and then grind with a pestle until finely divided. Add water in small portions to wash out the contents. Add the rest of the water and mix well.

11. Gram's Safranin

Safranin   0.25 g.
Ethanol (95%)   10.00 ml.
Distilled Water   100.00 ml.
Dissolve the safranin in the ethanol. Mix thoroughly. Add the distilled water and stir well. Then filter through filter paper.

12. Leifson's Flagella Stain

Flagella Stain
  (obtain commercially) 1.7 g.
Ethanol (95%)   35.0 ml.
Distilled Water   65.0 ml.
Dissolve the flagella stain in the ethanol. Add the distilled water. To completely dissolve the dye, shake the solution often during a 10 min. interval. The stain is stable for two weeks if kept tightly stoppered.

13. Loeffler's Methylene Blue

Methylene Blue   0.3 g.
Ethanol (95%)   30.0 ml.
Distilled Water   100.0 ml.
Dissolve the methylene blue in the ethanol. Add the distilled water and mix. Then filter through filter paper.

14. Lugol's Iodine Solution

Iodine, C.P.   50.0 g.
Potassium Iodide,
  C.P. (KI)   100.0 g.
Distilled Water   1000.0 ml.
Mix the iodine and potassium iodide in a mortar and triturate with a pestle until finely divided. Add distilled water in small portions to wash the contents into a beaker. Add the rest of the distilled water. Stir until completely mixed.

15. Malachite Green (5%)

Malachite Green   5.0 g.
Dissolve the malachite green in 100 ml. of distilled water.

16. Methylene Blue (0.5%)

Methylene Blue   1.25 g.
Dissolve the methylene blue in 250 ml. of distilled water.

17. Methylene Blue (1:10,000)

Methylene Blue   0.01 g.
Dissolve the methylene blue in 100 ml. of distilled water.

18. **Methyl Green (1%)**

    Methyl Green   1.0 g.
    Dissolve the methyl green in 100 ml. distilled water.

19. **Methyl Red Indicator Solution**

    Methyl Red   0.1 g.
    Ethanol (95%)   250.0 ml.
    Distilled Water   250.0 ml.
    Dissolve the methyl red in the ethanol. Add the distilled water. Mix well and filter through filter paper.

20. **Phenol Red Indicator Solution**

    Phenol Red   0.2 g.
    Ethanol (95%)   500.0 ml.
    Distilled Water   500.0 ml.
    Dissolve the phenol red in the ethanol. Add the distilled water and mix. Filter through filter paper.

21. **Safranin (0.5%)**

    Safranin   0.5 g.
    Dissolve the safranin in 100 ml. of distilled water.

22. **Sudan Black B**

    Sudan Black B   0.3 g.
    Ethanol (95%)   75.0 ml.
    Dissolve the sudan black B in the ethanol. Add 25 ml. of distilled water and mix thoroughly.

23. **Toluidine Blue (0.1%) in Ethanol (10%)**

    Toluidine Blue   0.5 g.
    Make up 500 ml. of 10% ethanol, then dissolve the toluidine blue in it.

24. **Wright's Stain and Buffer**

    Obtain commercially; "homemade" preparations are often faulty.

25. **Ziehl-Neelsen's Carbolfuchsin**

    Basic Fuchsin   0.3 g.
    Ethanol (95%)   10.0 ml.
    Phenol Crystals, C.P.   5.0 g.
    Distilled Water   95.0 ml.
    Dissolve the basic fuchsin in the ethanol. In another container dissolve the phenol crystals in the distilled water. Mix the two solutions.

APPENDIX E

**SUGGESTED SCHEDULE OF EXPERIMENTS FOR A
17-WEEK SEMESTER (Assuming Two Laboratory Periods Per Week)**

Week No.	Begin	Follow-up
1.		
a.	1.1	
b.	1.2, 1.3	
2.		
a.	1.4, 1.5, 2.1	1.3
b.	2.2, 2.3	1.4, 1.5
3.		
a.	2.4, 2.5	
b.	2.6	
4.		
a.	2.7, 2.8	
b.	2.9, 2.10	
5.		
a.	2.11, 2.12	
b.	2.13, 2.14	
6.		
a.	2.15	2.13
b.	3.1	2.15
7.		
a.	3.2, 3.3	3.1
b.	4.1, 4.2, 4.3	3.2, 3.3
8.		
a.	4.4, 4.5	4.1, 4.2, 4.3
b.	4.6, 4.7, 4.8	4.4

9.
- a.   4.9, 4.10, 4.11   4.6
- b.   4.12, 4.13   4.7, 4.8

10.
- a.   4.14, 4.15   4.10, 4.11
- b.   5.1, 5.2   4.12, 4.13

11.
- a.   5.3, 5.4   4.14, 4.15
- b.   5.5, 5.6, 5.7   5.1, 5.2

12.
- a.   5.8, 5.9   5.3, 5.4
- b.   5.10, 5.11   5.5, 5.6, 5.7

13.
- a.   6.1   5.8, 5.9
- b.   6.1   5.10, 5.11

14.
- a.   6.1, 7.1, 7.2   6.1
- b.   7.3, 7.4, 7.5   6.1, 7.1, 7.2

15.
- a.   7.6, 7.7, 7.8   7.3, 7.4, 7.5
- b.   8.1, 8.2   7.6, 7.7, 7.8

16.
- a.   8.1, 8.3, 8.4   8.1, 8.2
- b.   8.1, 9.1   8.3, 8.4

17.
- a.   9.2, 9.3   9.1
- b.   9.3, 9.4   9.2

APPENDIX F

_____ DAILY PLANNING CALENDAR
MONTH

APPENDIX F

_____ DAILY PLANNING CALENDAR
MONTH

_____ DAILY PLANNING CALENDAR
MONTH

APPENDIX F   311

_____   DAILY PLANNING CALENDAR
    MONTH

_____   DAILY PLANNING CALENDAR
    MONTH